First Edition October 25, 2013

Copyright © 2013 by Craig Whitman

All rights reserved. This book or any portion thereof
may not be reproduced or used in any manner whatsoever without the express written permission of the publisher except for the use of brief quotations in a book review.

Printed in the United States of America

Whitman Publishing
701 Colima Dr.
Toledo, OH 43609

.

In great appreciation for the support of my wife who had to put up with me through the entire process of completing my first book. Without her patience I would have never been able to accomplish this task.

This work began out of an acute interest in out of place artifacts and their implications for the human race along with the fact that mainstream archeology appears to have a lack of interest in these items or any hypothesis that does not fit the current and accepted view. Every effort at my disposal has been used to verify information used in this writing. This being my first attempt at writing there may be many mistakes contained in the work.

All of the photos used in this work were taken from the internet free use sites. Due to the multiple sources found for each photo it was not possible to determine who the true copyright owner might be so the "fair use" rule of copyright law has been assumed.

If you are the original owner of a photo that has been used in this work and have any objection to its use in context or in general or if the photo was on one of the free use sites by mistake contact me and I will happily work with you to arrange an agreeable solution.

All writing in this work is original. Facts and figures from hundreds of websites and books were used in the research needed to complete the work. Any resemblance to work by any one else is purely coincidental.

C.E. Whitman

Chapter 1

Modern Science and religion

Science and Religion have only recently achieved some semblance of equilibrium. Today's science should be far advanced compared to where we are at this time. Unfortunately the primary for cause for this dates back to the Roman Empire and the beginnings of a truly organized religion. During the latter half or the first millennium the church had gained great power over the populace. When the Empire collapsed the church gained control and began ruling the known world. Most science was viewed as heresy and was often punishable by death. It is well known that Giordano Bruno was burned at the stake in fifteen forty eight for his beliefs that the universe was full of life and the sun was just a star and other beliefs that the church considered to be contradictory to their doctrine. Any one that studies history with an open mind can easily see that the church held mankind back from progress for at least seven hundred years. Take that thought and try to imagine where science will go during the next seven hundred years and you will reach the minimum point where we should be now. This statement is not to criticize the church it is simply to state the fact and to point out that this kind of behavior is a human fallacy that we will probably always have to deal with or over come. This lack of progress could just have easily been caused by any other entity that might have gained power like the great library of Alexandria being burned by the Romans. The knowledge lost at that time might have advanced modern science as much as a thousand years or more. There is little doubt that everything there was to know about ancient Egypt and the great pyramids would have been contained in that library. The Egyptians knew how important knowledge was, that's why the great library was there in the first place. What a wonderful thing it would be to find a duplicate some where in the future.

Chapter 2

Fringe Archeology and the Sciences

Fringe Archeology often refereed to as Forbidden Archeology, pseudo science, and several other terms used by some people to describe interpretations that are outside the mainstream or Academic Archeology as well as the other sciences. Many theories developed this way are quite often radically different from Mainstream or Academic Archeology and Science. The Academic Sciences have heavily criticized the Fringe claiming that it relies on sensationalism, misuse of logic and evidence, and misunderstanding of scientific method. The mainstream apparently ignores the fact that they have been wrong every century so far. The eighteen hundreds re-wrote most of the accepted science of the seventeen hundreds and then in the nineteen hundreds re-wrote the accepted science of the eighteen hundreds. Just imagine what we will know at the end of the twenty first after so much of the current accepted science is re-written.

One of the most compelling theories in Fringe Archeology is the evidence that suggests that civilization has been around a lot longer than is believed by the Mainstream. Or the implication that many civilizations have evolved over time and were destroyed leaving very little evidence of their existence.

No Science progresses without new ideas and or new information. Quite often this new information or new idea comes from outside the mainstream community. Often discovered by accident these ideas and or discoveries once proven as authentic will initiate dramatic changes. Unusual theories will get little of any attention from the mainstream until the theory is proven as fact. However, once proven the mainstream will of course become very interested and seek as much credit for rewriting history as possible.

There are many different interpretations of the past that are at odds from those developed by academics (who often disagree among their own ranks). Most academics believe they have a fairly clear picture of the past until something shows up that proves they do not. However, one theory is as good as another until one or the other is proven to be fact.

When we visit a museum we see artifacts that have been arranged to support the current premise that man evolved from primitive creatures and has steadily progressed upwards towards technology and civilization. This view written by the mainstream is presented by most history books.

Fringe Archeology presents different views based on tantalizing objects (called erratics or out of place artifacts) that have been unearthed, which historians and academic archeology wish to ignore or quickly disprove. These objects (while problematic to the mainstream) suggest that man or possibly and unknown species very similar to modern man reached a level equal to or possibly beyond the current level of technology many times in the past and through some catastrophic event has digressed into a primitive state creating the necessity to start over. There are literally thousands of objects and sites that suggest this rise and fall of past civilization. Some interesting objects that question the accepted view are listed later in this writing.

At the beginning of the nineteenth century those who believed that the city of Troy actually existed would have been considered to be Fringe Archaeologists the city having been declared a myth by mainstream archaeologist. Those who did not conform to the scientific paradigm of the day were generally ridiculed by those of the mainstream. Ridicule is often the best way to silence an idea that does not conform to the accepted knowledge of the time.

Heinrich Schliemann was not an archaeologist when he set out to find the city of Troy; the science of archeology itself was still very young and had not developed into a true profession. But he was an avid believer that the works of Homer were based on real events and real locations. Although Schliemann worked several excavations the one being discussed here was his greatest find.

It was not until eighteen sixty eight that Schliemann arrived at Troy's location. Frank Calvert and/or his family owned the land and Calvert believed it was the true location of Troy. Being a very wealthy man Schliemann could afford to undertake the excavation and was convinced to do so by Calvert. Excavation began in eighteen seventy one and the second city of Troy, was found that same year. The dig was shut down by the Turkish government in eighteen seventy four and remained closed until it was reopened in late eighteen seventy eight or early seventy nine. Schliemann worked a total of four excavations at this site before his death in eighteen ninety. Although he was criticized for his use of dynamite in his excavations afterward he did find what he was looking for and by doing so actually pioneered the science of field archeology.

Wilhelm Doerpfeld who was an assistant to Schliemann continued the work after Mr. Schliemann's death and uncovered nine different periods of Troy's existence. Academic Archaeologists were now forced to accept that Troy was something other than a myth. This is only one of many examples of Fringe Archeology that later became mainstream science.

Mainstream science has often declared items that do not fit the current theories as either myth, impossible or simply fake, often without a thorough examination of the evidence.

Far too often it requires someone outside the mainstream to prove that something is real or that the conventional theory is not correct. The consequences of what these people find or prove often have a great impact on the world.

Chapter 3

If our current civilization collapsed tomorrow would there be evidence that we ever existed after a hundred thousand years, or a million? Probably very little evidence if any at all would be left. For one thing humans are really quite small and frail compared to the dinosaurs and technology doesn't really last very long. Another thing is population. Only two thousand years ago the world population is estimated to have only been about one hundred million. When the pyramids are believed to have been built the estimate shows less than fifteen million people world wide.

Proof of the short time it takes for humans and human endeavors to disappear from the record can easily bee seen through the tombs and monuments throughout the ancient world. Mummification has preserved some human and animal remains for a few thousand years but it is highly unlikely that even mummies will last for tens of thousands or millions of years. The only thing known that can last for so long is stone. Furthermore, even what we do find from the past few thousand years is usually buried and excavation is necessary to get to it. Needless to say the deeper we dig the more that is uncovered. It remains a point of interest that a period with so few people could build things that today can not be completely understood. One has to think they must have had some (possibly leftover) technology that has been lost between their time and ours.

The mainstream often claims to have an understanding of things (like the pyramids) when the truth is the understanding is cursory at best. They say today that the mysteries of the pyramids have been solved but the truth is the investigation is just beginning. We only need to see the recent news over the past few years to know they do not have a complete understanding of the structures. For example a few years ago a small robot was sent up a shaft that most Egyptologist said was just a ventilation shaft the robot found a door in the shaft and the exploration was halted at that time. Then recently another robot was sent up the same shaft and drilled a hole that a small snake camera could fit through and when the camera was inserted a new room in the structure was discovered. Once thought to be built as tombs (a supposition many in the mainstream still hold on to) the real purpose of these great structures has not truly been determined. No mummy has ever been found in one. Some controversial new research even suggests that they may have been colossal power generators.

There is abundant evidence of the dinosaurs because they were very large animals. Their bones would not decay quickly so they had a lot of time to become embedded in the material they were lying in when they died allowing them to become fossilized. We have evidence of smaller species down to microscopic life forms because they died in material that deprived them of the necessary ingredients needed to decay, this material later becoming rock, coal or quartz.

The earth has been around for about four billion years and yet we only see evidence from our species for the past (if we count Lucy) six million years or so. Most evidence of our species is from the past fifty thousand years with a sudden wide scale burst into civilization (until recently) at about five or six thousand years ago. It took a recent accidental discovery that the mainstream could not theorize away to more than double the accepted time frame for human civilization, that is to say evidence that mainstream science has now generally accepted. However, objects that defy explanation have been found all over the world for close to five hundred years and continue to be found today.

Lucy is one of the oldest known hominids and is believed to be about six million years old. Lucy is considered to be the beginning of the human race even though reconstructive modeling clearly shows that she was a lot more ape than human. The theory of evolution still has many missing links (the main reason that evolution is still considered a theory). Evolution may turn out to be fact at some point but it is most likely that if it does become fact it is going to date much further back than the current theory suggest. There have been other hominid species that lived on this planet in what we know of earth history. One or more of them may have developed civilizations that have become extinct. Leaving behind some of the ancient artifacts we find today.

The few things that are found that predate the accepted theory are rarely investigated thoroughly enough to determine the correct age or origin of the objects. Although there are many items today that were produced to mislead, there are many that are true finds that predate human activity. Indeed many artifacts have been found that were not understood when they were found and some that are not understood today.

There are artifacts that are extremely difficult to produce even with today's technology. And artifacts that the human race could not produce until many years after the items were found. It is remarkable that the mainstream will cling to the supposition that an article is not legitimate just because modern man learned to make a replica fifty to a hundred years after it was found.

It is also unlikely that identical technologies from prehistory would be reproduced with the rise of a new civilization. . A new civilization might develop something similar to a previous one using different methods of producing power and machinery. Just because our science has developed down a particular path does no mean that any other civilization would have chosen the same direction yet they may have obtained the same or similar results. It is unlikely that modern science would immediately understand a technology that advanced different than what is known today. Another species or ancient humans might not have even thought in terms that could be understood by modern humans.

Chapter 4

Since the dawn of human history man has wondered where he came from and how he got here. Answers to these questions have eluded the human race and still remain elusive today. Needless to say there are many different theories and avenues of research that have been and still are being pursued in the attempt to find out where the human race really came from and who we are. It is a common belief that only a very small piece of the puzzle presents itself.

The following information is an accumulation of some of the artifacts that have been found that suggest that just maybe mainstream science is missing more than a few pieces of the big picture. It would be desirable to get some people with the accredited skill sets (the mainstream) to investigate and at least try to form an honest opinion on these objects. Until that time arrives it is up to the reader to decide what is and is not a valid assumption.

It is almost inconceivable that Giants like the dinosaurs would develop before the smaller creatures and then have to die out so that smaller species and mammals could take over. After all even mainstream science agrees that all life begins as a single cell. Jumping from tiny to giant seems unlikely. Of course it could be that many of the smaller species including man lived along with the dinosaurs.

Although the known artifacts are limited and scattered through out the world but they are there. Much as some things from our civilization might survive for many millions of years but those items would be scattered and difficult to find or identify. The only material that would survive in abundance for such long periods of time would be very hard stone, crystal or granite. So to leave lasting records in other mediums would be at the least very difficult. Paper rots quickly and even our solid state technology in use today would probably not survive a thousand years.

We are only today learning to embed information in crystal which is predicted could hold that information for millions of years. A few short years from now we will most likely be using crystal for information storage the same way we use solid state drives today. The difference will be that the crystal will be able to hold millions of times the information and maintain the integrity of that information millions of years longer.

There have been many different theories and speculations presented concerning the source of out of place artifacts however a definitive explanation remains unproven and possibly unrepresented. This is partly because it is difficult for those outside the mainstream to get research funding and partly due to the fact that it is nearly impossible to convince someone in the mainstream to risk his or her credibility and pursue something as ridiculed as ancient artifacts and catastrophic theory.

As time passes more and more discoveries are popping up that change the understanding of human history. Recent (accidental) discoveries have pushed the archeological dating of the rise of human civilization back to almost double the previously accepted time line.

Even with recent discoveries mainstream archeologist are still reluctant to explore out of place artifacts. At some point someone will stumble upon something that provides irrefutable proof that civilization has rose and fell many times during the history of this planet. It is very difficult to even imagine what form that new information could come in but there is little doubt of the impact it would have on the modern world.

There are always debunkers when anything is discovered that does not have a conventional explanation. Mainstream science seems to have a tendency to tilt findings to conform to the current theories.

Then when there is a discovery that will not allow them to this they will make as minor an adjustment to the current theory as possible so that it does not offend their sensibilities.

There are also extremist beliefs such as religions, ancient alien theorist and so forth. That will propose the most imaginative theories possible. The religions will completely disregard any science not based on some form of creationism. On the other hand ancient alien theorist (while they may turn out to be right in the end) will have a tendency to incorporate many possibilities into alien intervention scenarios.

To gain some perspective on the subject of a different view of human history it is necessary to explore some of the possible causes for the fall of advanced civilizations that have may have existed (so far back that there is no history or accepted evidence) and disappeared. This is simply an attempt to explore another possibility and allow the readers to come to their own conclusions. No disrespect to any theory whether it was developed by mainstream science or any other source is intended.

Extinction Level Events

To begin with Yellow Stone has erupted several times during the course of earth history. Eruptions from volcanoes of this size are known as extinction level events meaning that most of the life on the planet would be destroyed within a short time of the eruption. If there was an advanced civilization prior to one of these eruptions there would be little evidence of that civilizations existence a short time after the eruption. Something like a recently discovered bridge (discussed later) could provide a clue as to the sophistication of a civilization that may have existed prior to one of these ELE's.

Yellowstone lies over a hot molten mantle which pushes rock toward the surface. Through a series of huge volcanic eruptions this mantel helped create the Snake River Plain to the west of Yellowstone. During the past 18 million years there have been a series of violent eruptions producing violent floods of basaltic lava. These eruptions have helped create the eastern part of the Snake River Plain which was once a mountainous region. At least a dozen of these eruptions were so massive that they are classified as super eruptions. Volcanic eruptions sometimes empty their stores of magma so swiftly that they cause the overlying land to collapse creating a great depression in the surface. The Yellowstone caldera which is one of if not the largest in the world was created when the lava from a volcanic eruption was blown out so quickly that it caused the surrounding land to collapse in this manor and created the depression which we see today.

Fortunately there are only six super calderas in the world. After Yellowstone we have the Long Valley Caldera, Valles Caldera,

Lake Toba, Taupo Caldera, and Aira Caldera. Should any one of these erupt it would not only change life as we know it but wipe most of it out. If anything or anyone survived we would most likely revert to a primitive state and again would have to rise from the ashes of our world to begin again. In addition to these super calderas we have many large volcanoes that could disrupt life on this planet on extreme levels. Some have cause mini ice ages during our own history. The best hope for our race at this time if one of these events occurred is our sheer numbers. With nearly seven billion people sever million at least should survive. Depending on who the survivors were would determine how far our world would digress.

In addition to large and super volcanoes we are also in jeopardy from objects from space. After all it is generally agreed that it was an asteroid that wiped out the dinosaurs. We are aware of several extinction level events caused by asteroids. It also appears that a major event happens about every fifty thousand years. A major event today could cause the world to revert back to a level many thousands of years past. In addition to the asteroids there are also comets that would wipe out most if not all life if they hit the earth. We also have the very real possibility of a global plague. We have seen several plagues that have had a devastating effect on the human race over the past thousand years. One recent example of this occurred in the early part of the twentieth century. The estimated population at the time was about one and a half billion. Five hundred million people or about one third of the world's population became infected with a killer flu between nineteen eighteen and nineteen. Estimated deaths from this pandemic ranged from twenty million to one hundred million. The first cases were seen in Europe but the infection spread around the entire world very quickly. Imagine an infection this deadly in today's world when one can travel around the world in a few days possibly spreading a new infection or disease to millions before the traveler is even aware he or she is sick.

Then we have the prospect of a massive solar eruption or solar flair that could literally cook the whole planet and destroy all life. Solar flairs occur all the time. So far the only massive ones that could have could have done great harm to the earth were blown out in other directions. However, we have had some midsized ones that reeked havoc on our power and communications systems.

There is no doubt there are countless unknowns that could accomplish the partial or destruction of the world we live in today. With so many threats there is little doubt that life has either been wiped out or at least crippled by these events many times during the past four billion years. There is also little doubt that our civilization is next in line for one of these catastrophes.

Without doubt the most dangerous of all to life on this planet is the human race. We have had the power to destroy life on this planet several times over for many years now. And the power to destroy is still increasing. There are laboratories around the world that have been experimenting with a huge assortment of deadly bacteria for decades. If only one strain of one of these bacteria got out to the population there would be little to nothing anyone could do to stop it. With our destructive nature we are the most likely short term threat to civilization. Even beyond our war like destructive capabilities we are responsible for pollution that increases daily, over population and over use of natural resources. While some of this is beginning to change as some of the world moves to a greener life style the danger is still there. The human race is responsible for the extinction of many species in our known history and in the end we may well be responsible for our own demise and possibly everything else on planet earth as well.

We need only look at the recent nuclear power plant disasters in America, Russia and Japan to see how easily our science and engineering to see how our modern world could be destroyed even without a war. When it comes to creating possibilities for disaster we humans are far ahead of anything nature has ever sent in our direction.

There may be a light at the end of the tunnel for our world though. For the past few decades science has been progressing speedily in directions that are not so destructive. Space exploration for one thing promises a future for the human race. As we become a space faring people we can insure that all knowledge gained during our time is not lost. We can also insure enough survivors that our way of life would not completely disappear. The possibility also exist that space exploration and colonization could solve enough of our issues at home that we could avoid our destruction and enjoy our world until the sun burns out a few billion years from now.

Medical science is also working on ways to extend life indefinitely. They are working on drugs that could render the population immune to infection from deadly viruses and possibly anything else. A breakthrough in any of the medical fields could easily insure that civilization would have the people plenty of people to rebuild after a catastrophic event.

Chapter 5

Chernobyl

On 26 April 1986 an explosion and fire at the Chernobyl power plant released large amounts of radioactive particles into the atmosphere. The contamination spread over most of the western part of the Soviet Union and Europe. Some where around a billion people have been exposed to excess radiation from this accident alone. Depending on whom one speaks with the eventual death toll from radiation related cancers ranges from a few thousand from the scientific community up to one million from the fringe community. Close to one half million people had to relocated from the area surrounding the plant. The immediate death toll from the accident ranges from thirty to sixty in and around the close vicinity. This has been classed as the worst nuclear in our history. Some photos were too graphic to include with this example. The ones that are included gives one a general idea of some of the after effects of this catastrophe.

After only twenty five years we can see that nature has already began to reclaim the area. However, clean up from this disaster will continue for many years to come and the damage to people and animals may last a lot longer than the cleanup.

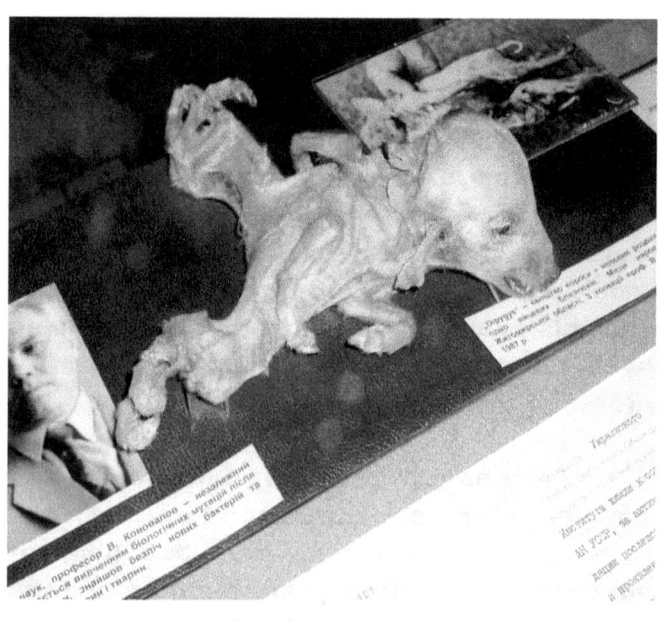

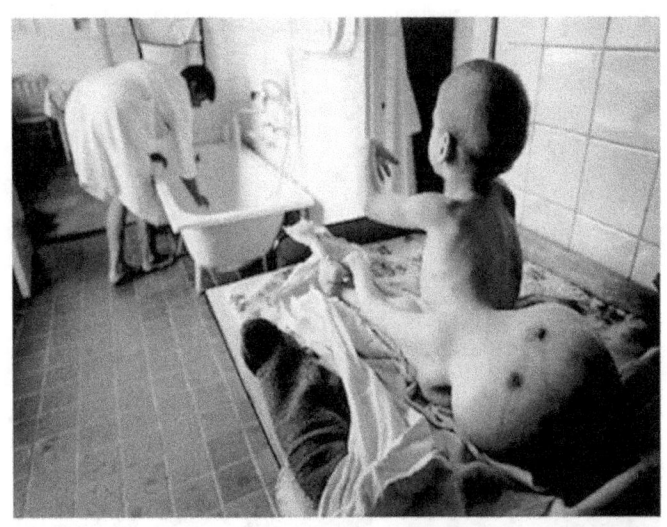

Fukushima, Japan

This is the only nuclear disaster caused by nature. This one started when a tsunami hit on March 11, 2011. The tsunami shut coolant control equipment resulting in meltdowns and releases of radioactive materials. The Fukushima is second only to Chernobyl with less than a third of the amount of material released.

Although over twenty thousand Japanese people died as a result of the earthquake and the resulting tsunami there have no reported deaths from radiation thanks to the quick response of the Japanese Government.

Three Mile Island

On March 28, 1979 the worst nuclear accident in Unites States history occurred at 3 Mile Island when a partial meltdown resulted in the release of small amounts of radioactive gases' and radioactive iodine. Failures in the secondary system and a stuck valve combined with inadequate training large amounts reactor coolant to escape. America was fortunate when this accident occurred. There were two million people living near this reactor and luckily they only received radiation amounting to about one third of a chest x-ray. However, an accident like this (uncontrolled) could easily wipe out a populated area this size or even larger.

Food and Water Contamination

As recently as 2011 according to the Centers for Disease Control close to twenty percent of America became sick from food contamination. About three thousand people died and somewhere around one hundred and twenty eight thousand were hospitalized. 128,000 were hospitalized. Bacteria, viruses, parasites, toxins and poisons, like poison mushrooms, are primary contributors to these events. There are currently five known pathogens that are responsible for the majority of deaths they are salmonella, toxoplasma, listeria, norovirus, and campylobacter sapp there are others that also contaminate food.

There have been at least ten major outbreaks of food borne illness in just the past one hundred years. This fact alone leaves little doubt that it will happen again. Unfortunately this is just an inevitable fact.

Since we ship various foods all from over the United States across the country and to other countries and acquire many from other countries the world population is always at risk.

If the water supplies of the Unites States were somehow compromised it could cause the death of millions very quickly.

Air pollution is still rampant around the world. During the late nineteen forties and nineteen fifties several thousand people died from localized air pollution. Larger events could spread far enough to kill millions if not billions in a very short time.

There are so many possibilities of human activities that could destroy life as we know it that it would be nearly impossible to list them all.

After a nuclear accident or explosion in any moderately advanced civilization it would not take long for the remnants of that civilization or the population of the effected area to vanish. In the aftermath disease would run rampant through the remaining population as well as possible starvation and the poisoning from the initial event. Scavengers would most likely remove anything thought to be useful, an action that would surely exacerbate the issue. Those that did not die would no doubt revert to a primitive state of mind focusing solely on survival. Birth defects would be rampant which could lead to the development of people that barely resembled those that existed prior to the event. Recovery from something like this could take hundreds if not thousands of years depending on available resources and possible aid in that recovery. We are aware of this due to witnessing such a scenario caused by our own actions. Only two atomic bombs have ever been used in our modern history both resulting in nightmarish consequences. Politicians made statements (after these bombs were used) vowing never again but kept funding research for and building more powerful weapons. It has taken a united effort for Japan to recover from the initial devastation but residual birth defects will most likely continue for many generations. If ten bombs had been dropped we would be looking at an entirely different story. There would probably be no Japan and the world would have lost one of its oldest and most honorable cultures.

Chapter 6

Evidence in India

In India there are several sites that appear to have been destroyed by tremendous heat. Rock walls appear to have been melted and fused together, much like the aftermath of a nuclear blast or a horrific nuclear accident. This work will only list the few that are best known.

The Mahabharata tells of a battle that describes an event that could resemble the aerial attacks that Japan suffered at the end of world two. Then (allowing for the vocabulary of the time) continues with a very accurate description of an atomic or nuclear blast and the subsequent consequences to the population.

> "Gurkha, flying a swift and powerful vimana hurled a single projectile Charged with all the power of the Universe. An incandescent column of smoke and flame As bright as the thousand suns rose in all its splendor, a perpendicular explosion with its billowing smoke clouds, the cloud of smoke rising after its first explosion formed into expanding round circles like the opening of giant parasols, it was an unknown weapon, An iron thunderbolt, A gigantic messenger of death,

Which reduced to ashes The entire race of the Vrishnis and the Andhakas. The corpses were so burned. As to be unrecognizable. The hair and nails fell out;
Pottery broke without apparent cause, And the birds turned white. After a few hours
all foodstuffs were infected. To escape from this fire the soldiers threw themselves in streams to wash themselves and their equipment. "

Mainstream science discounts the details that are described in the Mahabharata believing first of all that it is not possible for ancient man to have possessed the knowledge to create such weapons and machinery and then that if there was a nuclear war it would not be isolated to a few cities in India. They have completely disregarded the descriptive nature of the text describing these events.

When evidence such as this presents itself and there is no mainstream theory to explain it they create some of their own alternative theories to fit within the current paradigm. That is precisely what occurred when the evidence of nuclear explosions or reactions scattered all over the world was discovered. It would appear that for the most part mainstream science seems to be unwilling to even consider the possibility that ancient or prehistoric man could have advanced to a level of technology that rivals that of today.

In Rajasthan, India a three-square mile area of radioactive ash was found where a housing development was being built. High rates of birth defects and cancer in the area prompted the original investigation that uncovered this radioactive area. Investigators at Rajasthan have uncovered evidence of a possible atomic blast dating back eight to twelve thousand years. The blast or what ever the cause may have been destroyed most buildings and may have destroyed as many as a half-million people. The preceding quote from the Mahabharata (while it may have been written hundreds of years after the event) could very well be a detailed description of a battle that caused this catastrophic.

Even today there are stories that were handed down by word of mouth for centuries before being put into print and legends that have become known to be factual. Yet the mainstream is unwilling to accept the possibility that the same thing could have happened in the distant past.

When Harappa and Mohenjo-Daro were excavated they found skeletons of unburied people scattered around the city. Some were still holding hands. There were no signs of any battle fought at this location. However, the walls and foundations of the city were fused together suggesting intense heat and rapid cooling. No volcanic or meteoric evidence is present at this site leaving an atomic blast (or some other weapon of mass destruction) as a distinct possibility. Some speculation of a nuclear accident from an ancient power plant that has been theorized by some to have existed on a nearby mountain top has also been proposed as a possible cause. Radiation levels in the remains of these ancient people (that were excavated before the modern nuclear era began) rival those found in the world war two victims of Hiroshima and Nagasaki after atomic bombs were detonated there close to the end of world war two.

The high radiation level in the skeletons makes the carbon-dating somewhat suspect since radiation can cause the bones to appear several thousand years younger than they actually are. Even the carbon-dating showed the skeletons to be in excess of four thousand five hundred years old. It may also be possible that these ruins predate the younger dryas period (a sudden cold spell or mini ice age that occurred after the last ice age ended).

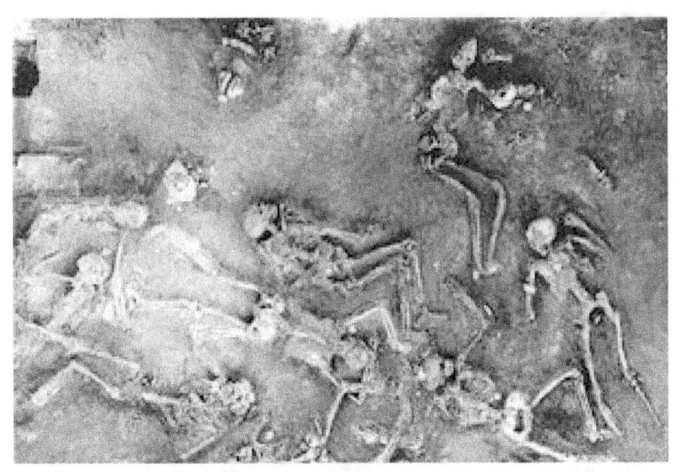

　　About two hundred and fifty miles from Bombay there is a crater known as the Lonar crater that is less than fifty thousand years old. There is evidence at this location that the crater was caused from pressure exceeding six hundred thousand atmospheres and extreme heat. The crater is nearly circular and six thousand feet in diameter. The crater is in the basalt about five hundred feet deep. Although no meteoric material has ever been found, ejecta from the blast can be found for nearly a mile around the site. Many of the ground zero sites of our own atomic testing have similar orators.

Lonar Crater

United States Nuclear Test Site

Chapter 7

Evidence around the World

The Gabon mine at Oklo in the African state of Gabon is the home of an ancient uranium mine. The Ancients could have mined the material to create their weapons here. There are also six of these ancient mines known in Libya and others that are scattered around the world. Moderated nuclear reactions have occurred in these locations producing plutonium. This information when first revealed created quite a stir in the scientific community. The very concept that some ancient people may have produced weapons grade uranium or created moderated nuclear reactions seems to insult many academics. It didn't take long for someone to come up with a half baked theory that some kind of earth pressure caused the reactions that depleted the uranium in these mines. Most of the mainstream was willing and eager to accept that theory or similar theories as they were presented. Going back to their accepted view as if the matter was settled seems to be the theme of the day.

At least 60 ancient forts throughout Scotland also show signs of what could be a nuclear blast. Complete with melted and fused stone. Another example of a possible

war using atomic weapons thousands of years before our recorded history began.

There is also evidence of some type of blast in many deserts around the world. The western United States has many ruins from some ancient catastrophe. Captain Ives William Walker was an American explorer and was the first to view and report many of these ruins. When Captain Walker first looked at the ruins in Death Valley seeing the wholesale destruction and the melted and fused rock he assumed that a volcano was responsible. However there has never been a volcano in this area. According to Captain Walker the area between the San Juan and Gila rivers is covered with ruins of cities, fused rock and craters. Obviously something happened here that created enormous heat possibly multiple atomic blasts. There is plenty of evidence in Death Valley that there were settlements thriving there in antiquity. This evidence doesn't seem to generate much interest for the mainstream.

It takes a temperature of about three thousand three hundred degrees to turn sand into glass yet we see large quantities of this greenish glass in nearly every desert in the world. Compare the glass from these deserts with the glass the atomic bomb left behind when the United States was testing atomic bombs at the White Sands missile range and we have an almost identical match. Falling meteorites might account for some desert glass but the glass created by a meteorite is usually murky and usually has a very low purity. Some scientist have made claims that the glass found in some of the deserts of Libya, the Sahara, Mojave as products of very large meteorite impacts however, there no craters have ever been found to back up those claims. Furthermore, the scientists in making the claims of meteoric impacts are completely ignoring the purity of the glass. It is simply not a plausible explanation for a meteorite impact to form glass that is close to ninety nine percent pure.

Glass from the Sahara

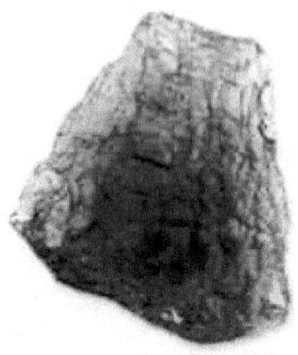

Glass from Modern Nuclear Testing

Chapter 8

Gobekli Tepe

One recent accidental discovery at Gobekli Tepe in 1994 turned out to be generally thought of as the greatest archaeological discovery in fifty years, but to many it is believed to be the greatest archaeological discovery in history. Prior to this discovery mainstream Archeology insisted that human civilization had only been around for fifty five to sixty five hundred years and that massive construction suddenly began about forty five hundred years ago . This single discovery with an estimated age of over thirteen thousand years more than doubled the mainstream timetable for civilization and virtually tripled the time frame for large construction projects.

We will have to wait to find out what the religious creationist will have to say about this discovery. They may have to readjust their time line for creation from six thousand years to thirteen or fourteen thousand. Or as with most new information that comes from artifacts or science that does not fit their beliefs they may just ignore it.

There was an archaeological survey of the area in nineteen sixty three by an American archaeologist and some of the tops of the stones were seen at that time but were assumed to be grave markers. This gave some room for the mainstream to quickly claim credit for the find.

Klaus Schmidt who is in charge of the excavation in conjunction with the Sanliurfa Museum of Turkey says the he decided to give the area a second look after reviewing the original survey. Another story told is that a local Sheppard discovered the site while attempting to move one of the stones. And while attempting to dig it out revealed some of the carvings and notified Sanliurfa Museum which is only ten miles away. Then the museum contacted Schmidt and sent him out to have a look.

Some of the stones in this structure are estimated to weigh as much as one hundred thousand pounds. The structures appear to range from thirty to one hundred feet in diameter. Pillars in the center are taller than the surrounding ones. This would logically lead the assumption that they once supported and intact roof.

Carvings on the stones show an assortment of wildlife lions, bulls, boars, foxes, gazelles, donkeys, reptiles including snakes, insects, arachnids, and birds, mostly vultures, some of the megaliths show crayfish or lions. Although difficult to imagine archaeologist believe that all of this stone was quarried, shaped, and decorated using tools made of flint. It seems to have escaped notice that the building and then back filling of this site correlates with the Younger Dryas period, the war that the Mahabharata speaks of and the dating of nuclear events around the world.

Photographs taken at Gobekli Tepe

David Lewis-Williams, professor of archeology at Witwatersrand University in Johannesburg, says: 'Gobekli Tepe is the most important archaeological site in the world.'

Some go even further and say the site and its implications are incredible. As Reading University professor Steve Mithen says: "Gobekli Tepe is too extraordinary for my mind to understand."

So what is it that has energized and astounded the sober world of academia?

The site of Gobekli Tepe is simple enough to describe. The oblong stones, unearthed by the shepherd, turned out to be the flat tops of these awesome, T-shaped megaliths. Imagine smoothly carved and slender versions of the stones of Avebury, Stonehenge or Easter Island. Most of these standing stones are inscribed with bizarre and delicate images.

Some of the stones seem to represent human forms - some have stylized 'arms', which angle down the sides. According to the archaeologist the site appears to be a temple, or ritual site, like the stone circles of Western Europe. However, there is no way for anyone to know for sure what these structures were at this time. They could have been dwellings since there so many of them.

After nearly twenty years of excavation less that ten percent of this site has been uncovered. The inner columns are arranged in circles from five to ten yards across. A lot of information has been deduced from this site but there are indications that much more are to come. Geomagnetic surveys imply that there are hundreds more standing stones, just waiting to be excavated.

Gobekli Tepe is currently considered to be the Garden of Eden come to life and possibly where the human story began. The first is its staggering age. Carbon-dating shows that the complex is at least thirteen thousand years old, possibly much older. That means it is believed to have been built around eleven thousand BC. Stonehenge was built in nine thousand years later around three thousand BC and the pyramids of Giza over nine thousand five hundred years later or about twenty five hundred BC. Gobekli is therefore the oldest know large construction site in the world, by a massive nine thousand year margin at least according to the mainstream. It is so old that it predates settled human life according to mainstream archeology. It is before pottery, before writing, before everything we have been taught by the mainstream. Gobekli was built during an era human history that for most people is unimaginably distant, right back to the time that the mainstream has always classified as our hunter-gatherer past. This is a period in history when people did not know how to grow food. Keep in mind that all of this was built prior to the point that we humans could even begin to estimate the world population. The estimated population for five thousand BC stands at five million and the structures here were built at least another six thousand years before that.

Judging by the rate of increase in the world population clock there would have less than a half million people at the time Gobekli was built. That's the mainstream version; the fringe version would be that our unknown ancestors from one of the previous civilizations built it.

How could it even be possible for cavemen to build something so ambitious? The mainstream speculates that bands of hunters would have gathered sporadically at the site, through the decades of construction, living in animal-skin tents, slaughtering local game for food.

They refer to the many flint arrowheads found around Gobekli to support this thesis because they also support their dating of the site. If one looks most anywhere in America arrow heads can be found. Does that then support the theory that everyone living in America is living in the Stone Age.

The speculation that there was a stationary population that built Gobekli in this time frame would be world changing in its own right. The possibility that Stone Age hunter-gatherers (generally nomads) could have built this place would be nothing short of phenomenal. Using their own population measurements alone should prove that there just were not enough people on the planet to build this site if any of their estimates are right. Again we run in to a scientific anomaly. People in this region of Turkey, were obviously far more advanced and sophisticated than the mainstream has ever conceived.

Archaeologists believe that Gobekli was buried intentionally to preserve the site. Considering the great amount of manpower and time it would take to bury this enormous site it should follow that the people there may have known that a major catastrophe (perhaps war) was eminent and therefore buried the site to save it for future generations to uncover. The mainstream will not like that kind of proposal but the theory should be as sound as simple abandonment as they have proposed.

It is almost unbelievable that even faced with such new information many in the mainstream still cling to the notion that the people that built Gobekli Tepe and then apparently covered it up were primitive hunter-gatherers. A project like this would take large work crews first to build and then to cover it up with sand. This is surely not an action that a small band of hunter-gatherers would have the manpower to accomplish. This can be considered a prime example of how difficult it is for the mainstream to let go of old ideas or incorporate new information into old ideas in order not to change.

Chapter 9

Underwater Monument

The Yonaguni Monument is a structure which was discovered by a diver off the southern coast of Japan near the island of Yonaguni Jima in 1986. This structure has many right angles, is six hundred feet wide and ninety feet tall with five separate levels of stone block and a road surrounding the structure. Yet without ever going to the site or diving on the structure many in the mainstream have labeled it as a natural formation despite the fact that many ancient tools were found around the structure.

These images Were Taken during Subsequent Dives

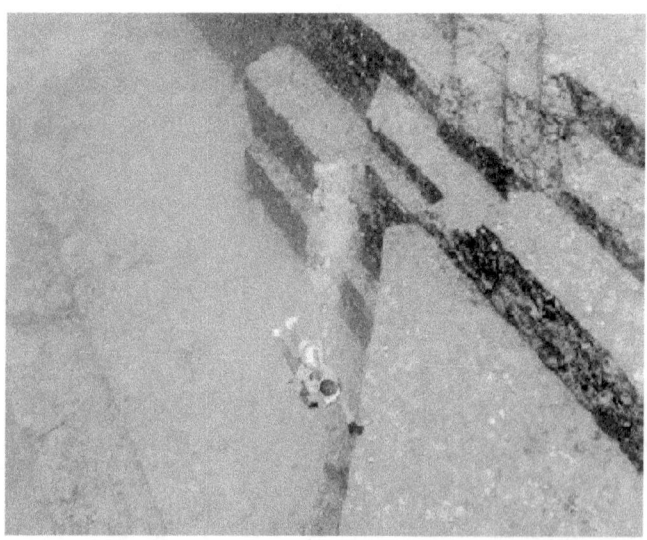

Standing near by the pyramid structure, what appears to be the carving of a human head was discovered along with numerous unknown hieroglyphs. The head standing several feet tall like the main structure has been labeled as a natural formation by many mainstream archeologists.

Although none of them have been able to explain away the hieroglyphs.

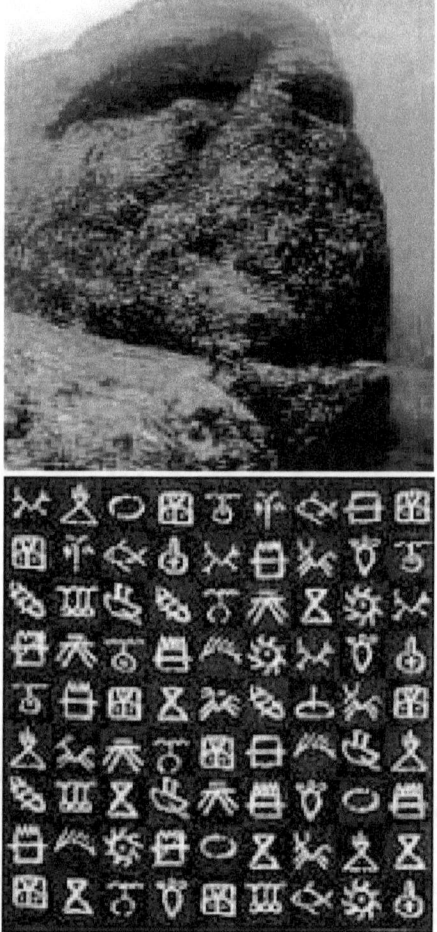

Chapter 10

Ancient Computer

An object was found in nineteen hundred while diving on a shipwreck in the Greek Islands known as the Antikythera

Device. It took nearly a hundred years to figure out what it was and some of its actual uses may still be unknown. The device was named after a nearby Greek island.

This device is nothing less than an analog computer. It is over two thousand years old and is believed (by most scientist) to have come from Corinth or possibly one of the Corinthian Colonies because that is where the ship came from. Another theory based on the language of the inscriptions which are in Koiné Greek (the language used throughout the Hellenistic period and spread widely during the conquest by Alexander the Great) leading the mainstream to the conclusion that this device was at least made somewhere in the Greek speaking world. Since the ship that was carrying this object went down around eighty BC there is nothing that tells for sure who built this device but it does appear to be based on Greek theories of astronomy and mathematics. The presumed date of the shipwreck, gear settings and inscriptions on the mechanism's faces were used for dating the device's construction. It may well have been a find from a much older culture or built from designs obtained at The Library of Pergamon or The Ancient Library of Alexandria which was only superior library in the world and supposedly held all knowledge gained throughout the known world. There is no doubt that there were predecessors to this device the accuracy of the device alone would dictate that this was not the first attempt to create something this complex. It took over a thousand years for modern civilization to even match the accuracy and complexity of the gears contained inside. Nothing in history even comes close until the fourteenth century when an astronomical clock was built.

Knowing the earth was a sphere circling the sun at that time is a phenomenal development considering the fact that even in the time of Christopher Columbus most people believed the world to be flat and that the earth was the center of the universe with everything else orbiting around it.

X-rays of the device have revealed some surprising functions. It is a mechanical computer of an accuracy thought impossible when the ship that carried it sank. Its wheels and gears create a portable orrery of the sky that predicted star and planet locations as well as lunar and solar eclipses and yet some have compared this complexity to that of a good quality eighteenth century watch. The Antikythera mechanism (shown here) is about thirteen inches high and similar in size to a large book. X-rays and advanced photography have uncovered its true complexity. This device is so remarkable that its discovery could be considered one of the greatest archaeological finds of all time. It also appears that what was recovered may not have the complete device. Some parts may have deteriorated beyond recognition or completely decayed with so much time in the ocean.

If we only have a part of the original it is not currently possible to state or ascertain all of the functions of the device. Perhaps someday as our science progresses someone may be able to determine what the possible missing parts were and reconstruct the device in its entirety.

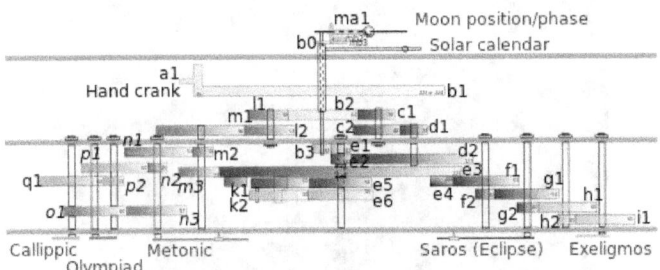

Chapter 11

Crystal Skulls

One of the most controversial items in existence today is known as the Mitchell-Hedges Crystal Skull. There have been many in the mainstream who tried valiantly to discredit this object and Mitchell Hedges, claiming that he purchased the skull at an auction which appears may be a half truth. According to his daughter, Mr. Hedges entrusted the skull to Sydney Burney as collateral on a loan. Mr. Burney put the skull in an auction and when Mr. Hedges found out about the auction he attended the auction and either by paying Mr. Burney or simply bidding at the auction he retrieved the skull however there is no known record of the loan.

This skull has been laboratory tested and found to be from a single piece of quartz which is nearly impossible to carve with such detail. It has no tool marks. The scientist who examined this item in the nineteen nineties were dumbfounded and stated that it would take several lifetimes working around the clock to carve this item with sand and even then there would be grooves in the material. Note that no groves were found even under a microscope.

Although there are inconsistencies surrounding the history of this incredible object the fact is that the technology to reproduce a skull even close to the complexity of this one did not exist until more than seventy years after the Mitchell Hedges discovery. And even the ones that are made today would not be able to stand up against all the testing that the Mitchell-Hedges Crystal Skull has been subjected to. This is a fact that the mainstream has completely ignored. There seems to some controversy over the origin and discovery of most of the known ancient crystal skulls. It really is amazing that the mainstream can so quickly label something as fake that still can not be reproduced with the accuracy or quality with current technology.

There are many legends surrounding these objects. Some claim near magical powers. Others speak of great knowledge. Even though current science has not been able to discern any embedded information from the skulls that does not mean that it is not there. Simply because skulls are made through out the world today the mainstream has labeled them all as fake. Again this was done without a proper comparison or any serious further investigation.

Just as we are beginning to embed information in crystal, an ancient civilization may have done the same or may have been further advanced in the process of storing their information in crystal. Science today is not able to detect information in these objects. Perhaps some future breakthrough might reveal that information is stored in these skulls and we will learn how to access it.

In most cases when science finds out that something exist such as an unknown language a way is found to understand it.

Learning how to ask the right question is always a good place to start. New testing should be considered with today's technology to see if we can learn more about the origin and the true purpose of these objects.

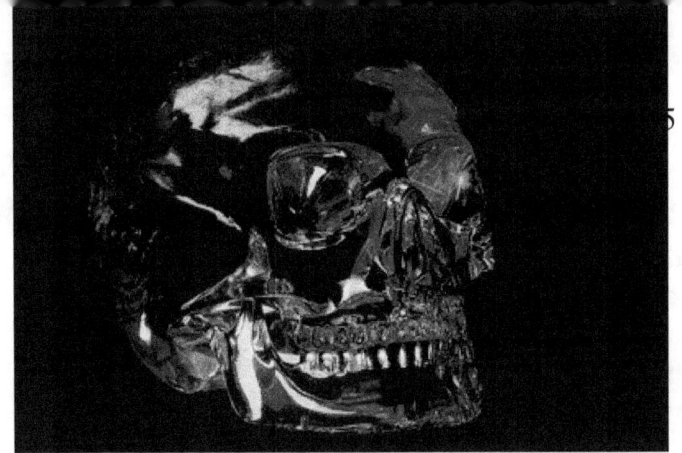

As of this date twelve are believed to have been found. According to most legends there are supposed to be thirteen. These legends also claim that when all thirteen are together the information contained within them will present itself beginning a new era of human history. Who knows? With so many ideas surrounding these remarkable items anything could be possible.

Perhaps our ancient ancestors left information about their civilization and what happened to them. If the civilization of the people who crafted these skulls faced the same or similar growing pains as our current one does that information could be of great benefit to our civilization. After all don't we try to learn from our own history?

The Amethyst Skull

The Amethyst Skull also known as Ami was discovered in the early 1900s in South America and was later brought to the United States. It is believed to have been owned by José de la Cruz Porfirio Díaz who was the Mexican President during the last twenty five years of the eighteen hundreds until nineteen eleven when he was ousted by the Mexican revolution. This skull is made of very dark purple amethyst. It has holes where the jaw meets the upper face, and circular indentations around the ears.

Like the Mitchell-Hedges skull, this one was studied at Hewlett-Packard, and was also found to be inexplicably cut against the axis of the crystal.

Sometime around nineteen seventy nine Ami came to the United States and shortly there after was used as collateral for a loan and later obtained by a group of businessmen who kept it in a vault until two thousand and nine when Ami again changed hands. Hopefully the new owner will see fit to share this treasure with the world.

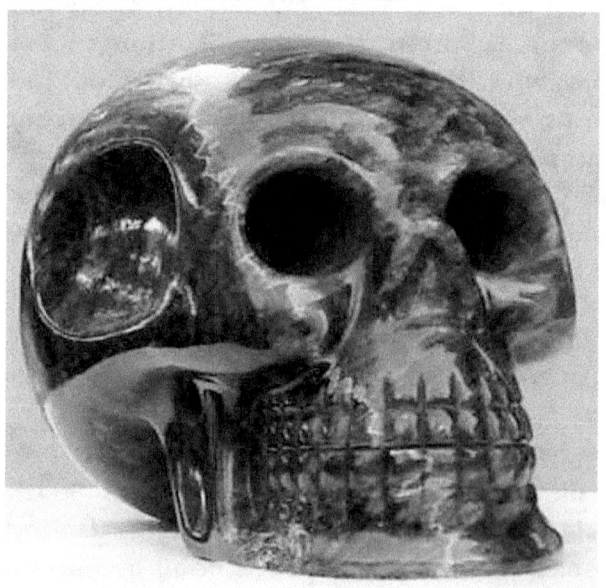

Texas Crystal Skull

The Texas Crystal Skull also known as "Max," is cut from a single-piece, clear quartz and weighs 18 pounds. It reportedly originated in Guatemala and was used by Mayan priests for healing, rituals, and prayers. The Mayan priests gave it to a Tibetan spiritualist who later passed on to Jo Ann Parks of Houston, Texas. The Parks family allows visitors to observe Max and they display the skull at various exhibitions across the United States. Max was examined and authenticated as well as possible by the British Museum.

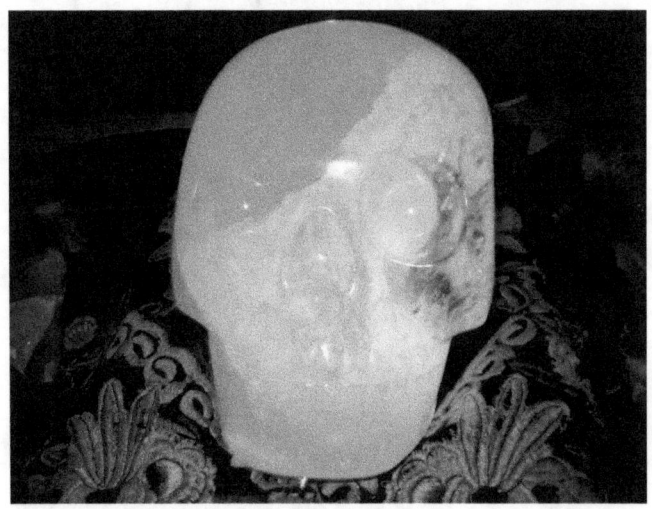

ET Skull

"ET" is a smoky quartz skull it was discovered in Central America by a Mayan family while digging on their property during the first decade of the twentieth Century. This skull was given its nickname because its elongated cranium and exaggerated overbite make it look like the skull of an alien being. ET is part of the private collection of Joke Van Dietan, who tours with her skulls to share the healing powers she believes they possess.

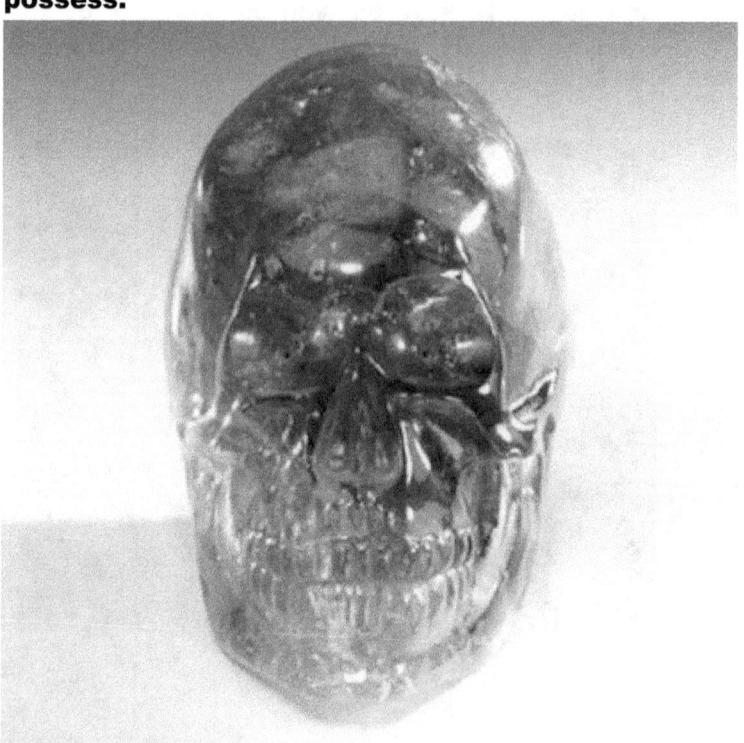

SHA NA RA Crystal Skull

SHA NA RA was found during an actual archeological dig in Mexico. This skull was also examined and authenticated by the British Museum. As with the other skulls that have been deemed to be truly authentic no tool marks were found on this one.

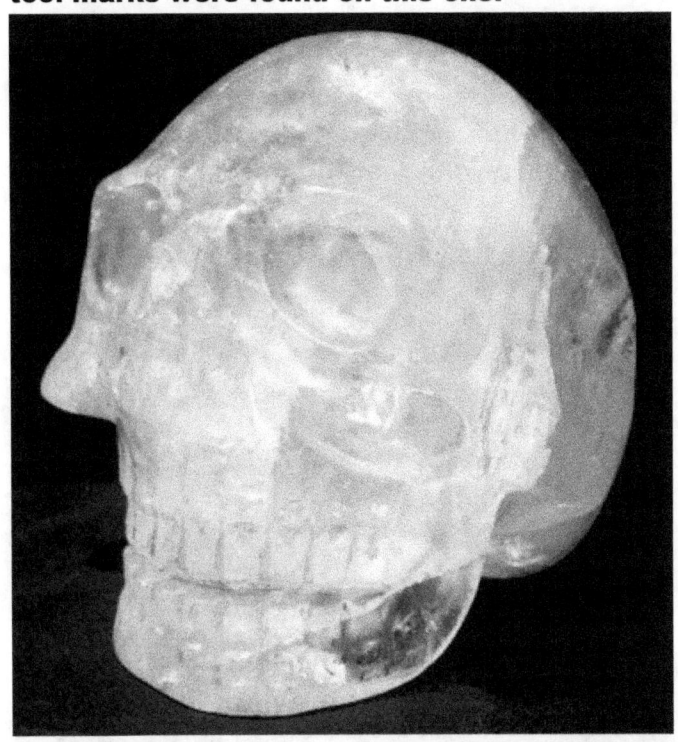

Mahasamatman Crystal Skull

Originally given the name of an Indian Prince this skull is now more commonly known as Sammie Girl. This is one of the

lighter skulls weighing only 4.41 pounds. Sammie Girl is not as well known as most other crystal skulls. Sammie Girl was taken to Damian Quinn (a leading expert in modern Brazilian hand carved crystal skulls) who authenticated this as ancient as well as possible. This skull is currently held by Kathleen Murray native to Scotland and appears to be residing in England. Note it is very difficult to authenticate a skull as ancient when science to not have a proven method or a valid test for authentication.

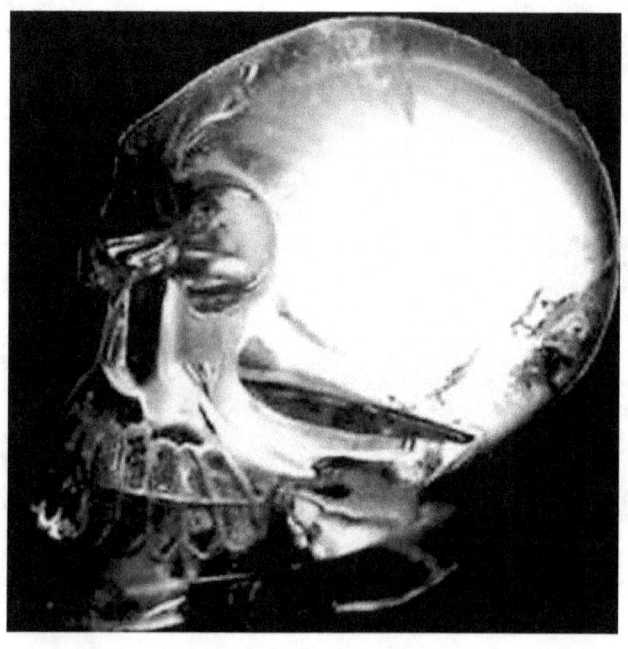

MAYAN CRYSTAL SKULL

The current location of the Mayan Crystal Skull is unknown at this time. There are two different stories concerning its discovery. The first story of says it was discovered in San Augustine, Guatemala in 1912. The second says that it was it was found at the Mayan site of Copan in Honduras in 1910. It appears to have come to the United States at the same time as the Amethyst Crystal Skull. The Mayan Crystal Skull is made of clear quartz and weighs 8.71 pounds. Again it is necessary to remember that there was no one known to be making crystal skulls at the time these were being found.

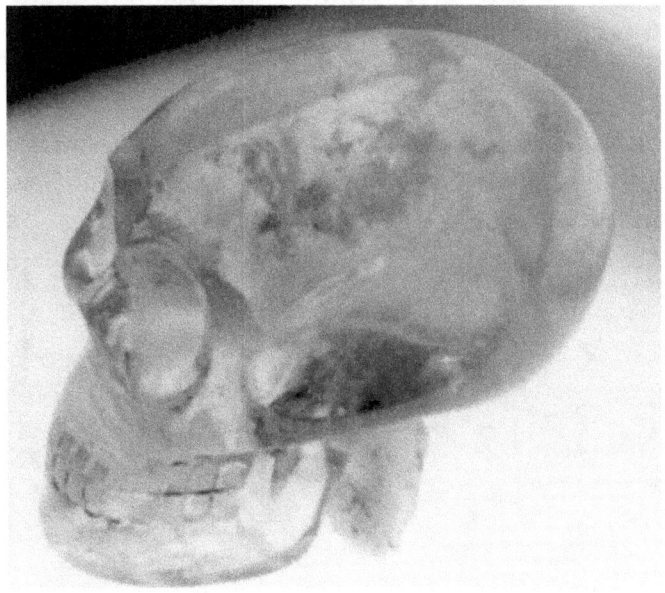

Crystal Skull Synergy

This skull was given to a hiker by an elderly gentleman living in a small village in nineteen eighty one. The hiker had stopped in the village to spend the night as he was hiking through the Andes, near the borders of Peru. He was told that it had been handed down from an old Nun that died in the early seventeen hundreds. That would have to mean that this skull has been around at least since the seventeenth century. Islanders claim to have sent the skull across the ocean centuries ago.

A forensic scientist examined the skull and stated that it had been modeled from a real face. He further explained that human skulls have abnormalities that only occur naturally and that artist conceptions do not reproduce these abnormalities. Sherry Whitfield Merrell is the current owner of this skull. She often travels with it and makes it available to the public.

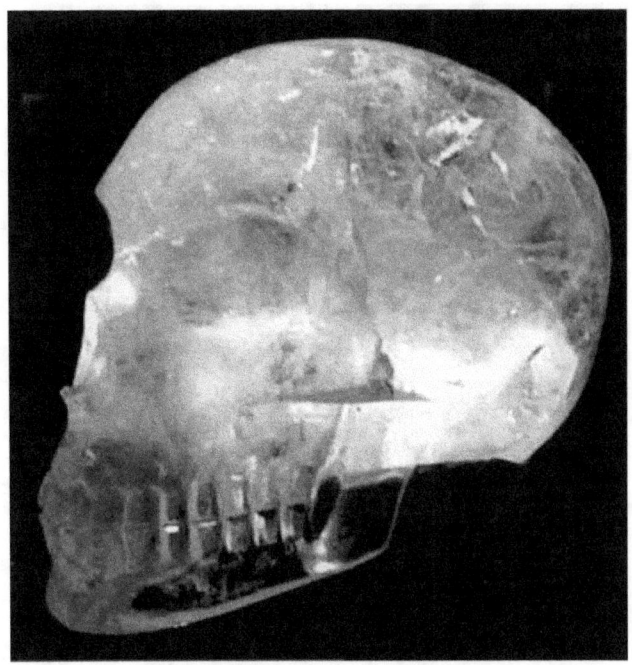

Einstein Crystal Skull

This skull was examined by Nick Nocerino (Known as the Father of Crystal Skull Research) one of the world's leading experts on crystal skulls. He stated that it was the largest and most anatomically correct of all the skulls he had examined.

Einstein is carved from milky quartz and weighs thirty three pounds. This skull is reputed to have brought to the United States during the nineteen twenties an explorer. Carolyn Ford obtained it indirectly from the estate of the explorer in nineteen eighty nine. There is currently no other information available about the origin of this skull.

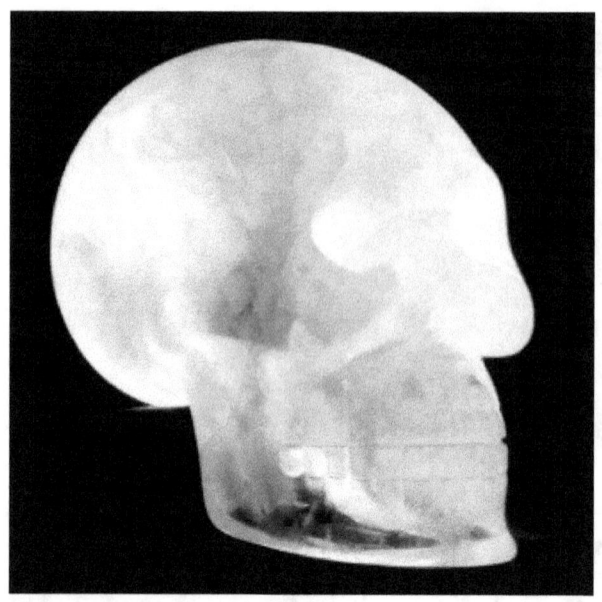

Chapter 12

Ancient Aircraft

During excavations in the central and coastal regions of South America artifacts were found that truly excite the imagination. Originally these items were thought to be representations of some type of insect. Airplanes were still a relatively new invention at the time.

During the latter half of the twentieth century working scale models built were using the designs and dimensions of these objects and they all flew perfectly.

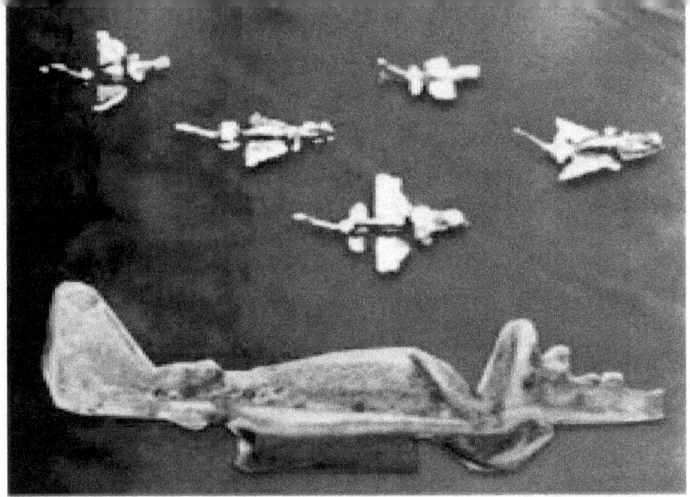

In addition we see that these models have a delta wing that allows for increased stability and maneuverability during flight an item that was not added to modern aircraft until the nineteen sixties. Other artifacts from the tomb indicate that the people who made these did use their imagination when making models of many things adding extra fins to fish and so forth. However the Ancient Incas were apparently aware of some serious aerodynamic design features. It would be highly doubtful that such a modern design would and such great detail would develop simply by an accident of the imagination.

Skeptics can not even consent to the idea that these ancient prehistoric people might have seen or even built machines of such complex design.

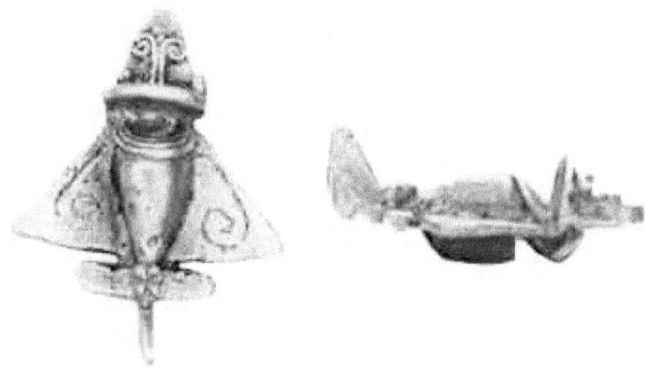

Somewhere buried beneath the sand or perhaps in a long forgotten cavern someone my very well stumble upon the remains of some ancient technology or something carved in stone that will explain these fantastic models. Until that time we can only guess as to the origin of the design for artifacts such as these. The most important thing to remember is that these air plane models were found in a tomb that was sealed many centuries before modern humans could fly.

Virtually every ancient culture around the world has legendary accounts of people who came from the sky. Suppose for a moment that after a great war that destroyed most of the planets population only a few remained that understood the technology. Wouldn't they spread out and try to rebuild the world? Of course if they had aircraft they would utilize that aircraft to move around the planet. The population that remained of course would accept the help from these people and elevate them to high status. Give it a few generations to progress and those descendants might be likely view the first generation as gods partly because they could fly and partly because they had been responsible for survival of the species. A scenario something like this could explain things like the golden planes pictured above and many other unusual artifacts around the world. It might also explain the technology used to build pyramids in different parts of the world. If there were not enough of these people to sufficiently reeducate those in the area they chose the technology and knowledge they were attempting to pass down could easily become lost in just a few generations. Much the way the technology the ancient Egyptians use to build their great monuments and civilization has been lost.

They Mayans

Another good example of ancient drawings would be the Mayan sarcophagus that appears to depict King Pakal flying in a rocket. Though highly disputed by the mainstream the image on this sarcophagus lid does appear to show the king in a vehicle.

As our civilization moves into the twenty first century our ability to find remnants of civilization's long past is improving. Hopefully some day soon some of the new methods being used will uncover records of some of the lost civilizations and prove once and for all that man has reached higher levels of civilization and technology than most people can even imagine. When this happens (and it will) the world will be changed forever. Such a discovery could take the human race in completely different directions and offer insights into many issues of our time. Understandably this is a highly optimistic view. Then the pessimistic view of such events would be more than most would like to see in print.

Even with all the technology at our disposal today we would not be capable of producing something like the Great Pyramid of Giza. We build enormous skyscrapers and send men into space but we haven't built anything that would stand thousands of years and still be as imposing as the things Egypt built more that four thousand years ago

Chapter 13

Nazca Lines

The Nazca Lines are a series of geoglyphs located in the desert of southern Peru. Although the mainstream dates the creation of these glyphs to between five hundred AD to five hundred BC they may be much older. In order to date the Nazca lines scientist used other art forms believed to have existed at that time. Using art form as a premise to date something like this is questionable to say the least. As with the current world art forms and other things have a tendency to repeat over time. So it is possible that we are looking of a repetitive art form when we look at the effigies created here.

The designs in this desert may have been something completely different than the many theories that have been proposed. It may be possible that these people were visited by more advanced people of their time and these lines were created in the hope that those (more advanced) people would return. His could be something like the cargo cults that sprang up in many villages in the South pacific at the end of World War Two. The cargo cults (as the were called) built effigies and piers and carved out landing strips in their fields hoping to entice the Americans that had landed there during the war to come back. Until the Americans landed there these Stone Age people had never seen an airplane up close and when one landed they believed the occupants to be some kind of gods. At the end of the war the Americans left. The indigenous people began to pray for the return of the ships and planes that had brought them so many wonderful things such as radios, canned meats, candy, and an assortment of modern conveniences. The islanders believed that these things and been produced by magic and sent to them from the spirit world.

Perhaps the people of Nazca were visited by someone flying something similar to the planes found in South America and were simply making the attempt to entice them back. The fact that our modern civilization coexists with stone age cultures that are scattered all over the world should actually serve as a wake up call to science an at least cause the mainstream to examine the possibility that the same thing could have happened many thousands and maybe even many millions of years in the past.

It is a common practice in aviation to use land marks as a navigational guide when flying under VFR rules. Theses could possibly have been placed in their location for the same purpose many thousands of years in the past. From the air some of this region does look like an airport. When we consider what we do not know about human history compared to what we do know it is amazing that mainstream science does not have a broader interest and a more open mind in attempting to solve riddles such as this that the ancients have left behind. Science is supposed a grand search for knowledge and enlightenment. The source to begin the search for that knowledge should not be limited to a particular paradigm or standing by the accepted view just because it is easier to conform than to question that view and take the chance that just maybe it is in error.

How can modern man be so smug as to assume that just because we have no written records of who created these effigies that they have no meaning or information that could be pertinent to our modern civilization. Such assumptions display a very narrow view especially for a species that often prides itself on insatiable curiosity.

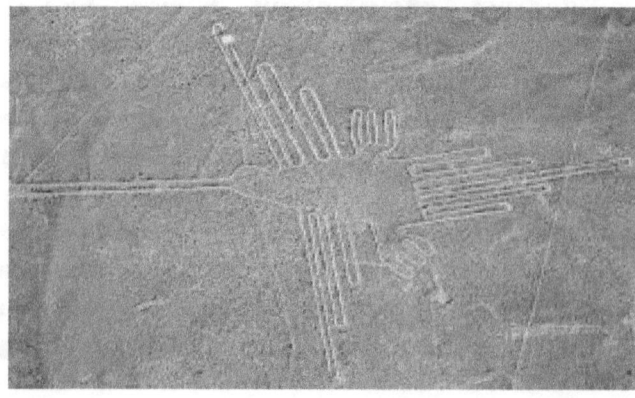

Chapter 14

Adam's Bridge

This image taken by NASA reveals a mysterious ancient bridge in the Palk Strait between India and Sri Lanka. Currently named Adam's Bridge it is made of a chain of shoals eighteen miles long.

The bridge's unique curvature and composition by age show that it is man made. The legends as well as Archeological studies reveal that the first signs of human inhabitants in Sri Lanka date back to the a primitive age, about one million seven hundred fifty thousand years ago and the bridge's age is also almost equivalent. According to the mainstream we were still at least one third ape at that time. Does anyone really believe that cave men or half apes built a bridge like this? This bridge should not exist according to the mainstream. Their current stance leaves them no choice except to come up with an explanation that fits into the current premise that defines the slow and steady rise of man. We must assume that such an explanation is forthcoming since it will take some time to find a way to explain this find to ac academic satisfaction.

This information is a crucial aspect for an insight into the mysterious legend called

Ramayana, which was supposed to have taken place in Treta Yuga (more than one million seven hundred thousand years ago).

In the Treta Yuga there are passages which speak of a bridge, which was built between Rameshwaram (India) and Sri Lankan coast under the supervision of a dynamic and invincible figure called Rama who is supposed to be the incarnation of the supreme.

This information may not be of much importance to the archeologists who are interested in exploring the origins of man or anthropologist that still search for fossilized records of missing inks in human evolution, but it is sure to interest the people of the world that have come to know of an ancient history linked to the Indian mythology that predates that of the mainstream.

Mainstream science agrees that the first signs of human inhabitants in Sri Lanka date back to a time before the Stone Age, about one and three fourths of a million years ago. Those people are believed to have come from the South of India and reached the Island through a land bridge connecting the Indian subcontinent to Sri Lanka named Adam's Bridge. This is related to be so in the epic Hindu book of Ramayana.

Also known as Rama's Bridge, this eighteen mile long chain of shoals is covered by four feet of water at high tide. Today a steamer ferry links Rameswaram, India, with Mannar, Sri Lanka. According to Hindu legend, the bridge was built to transport

Rama (hero of the Ramayana) to the island to rescue his wife Sita from the demon king Ravana. New discoveries and accidental under water finds may soon give way to our understanding of this ancient culture as well as the time line for the history of our species and the world.

Here is a better look from Google

The recent discovery of a sunken city off the Northwest Coast of India near Surat could revolutionize our concept of history. All the history textbooks would have to be rewritten if this ancient find proves to be of Vedic origin. Murli Monohar Joshi, Minister for Science and Technology in India confirmed the archeological find by an Indian oceanographic survey team. Radiocarbon testing of a piece of wood from the underwater site shows an age of 9,500 years, making it four thousand years older than earliest cities currently known except for Gobekli Tepe.

The ancient Sanskrit writings of India speak of cities existing on the Indian subcontinent in very ancient times. Although most historians consider these accounts as mythological, new discoveries promise to confirm some of the old literary accounts.

Most historians believe that Sanskrit-speaking people migrated to the Indian subcontinent about thirty five hundred years ago, from Central Asia. However there some who accept that India itself is the original home of these people. Much the way Christians today use the bible to guide their lives, these ancient people used a body of literature called the Vedas. For this reason their culture was known as Vedic.

It is unfortunate that the accepted history of India was written mostly by westerners with very little regard for the ten or twelve thousand years of written material that India has preserved. Modern historians and scientist regard the records that India has as mythological. Indeed some of the ancient records do read like modern science fiction but then just one hundred years ago a man traveling to the moon in a rocket ship or flying through the air in a plane was fanciful science fiction. When we look around our modern world the technology of today is far beyond anything that was imagined at the beginning of the last century (with the exception of a few science fiction writers). Mainstream science has become a little more liberal today than they were a few years ago with many of today's finest science fiction authors being mainstream scientist and a lot more open minded that those of the past. This is most likely due to the huge number of times that science has been mistaken when declaring something impossible or their attempts to visualize the future.

    The ancient texts not only speak of many different flying machines called vimana, many also have detailed descriptions of these machines. Some artist depictions of these descriptions found on the internet are posted here. It must be remembered that when these were drawn we had not yet developed any of the technology necessary to make any of them fly. The Shakuna Vimana would appear to be using rocket power possibly combined with (what is known today as) a ramjet. As a point of interest consider that there are literally hundreds of varying descriptions of these and other flying machines in these ancient text. It is absolutely amazing that people considered by the mainstream to have lived before civilization even began would be able to come up with design detail that rival today's engineering capacity.

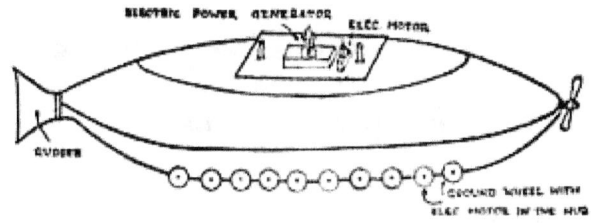

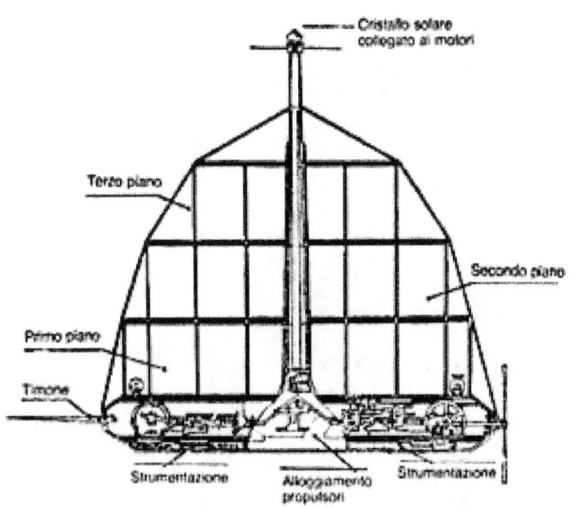

SHAKUNA VIMANA
HORIZONTAL SECTION

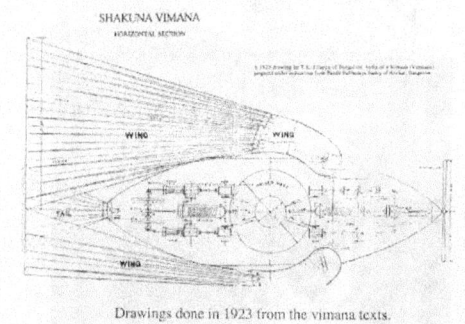

Drawings done in 1923 from the vimana texts.

To gain some extra perspective we can observe the similarities between Illustrations 2 and 3 with our own Gemini Capsule.

RUKMA VIMANA

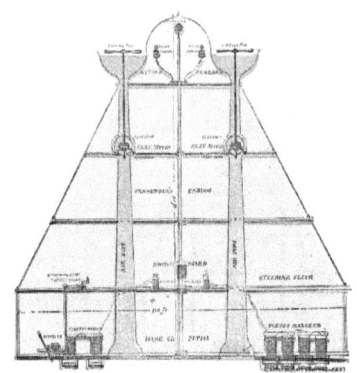

VERTICAL SECTION

Drawn by
T. K. ELLAPPA,
Bangalore.
2-12-1923.

Prepared under instruction of
Pandit SUBBARAYA SASTRY,
of Anekal, Bangalore

There are many cold periods and Ice Ages in the history of this planet the most recent three in prehistory are known as the Younger Dryas, Older Dryas, and the Oldest Dryas these were interruptions of the global warming from the last Ice age.. The younger dryas began about twelve thousand nine hundred years ago and lasted for about three thousand years. The Older dryas period ends with a sudden drop in temperatures on a global scale. The planet had been on a steady warming trend for over a thousand years at the time, beginning with the end of the last ice age. There are several theories about the cause of this cold spell including impact scenarios that really have no foundation. The fact that the earth cooled very rapidly is the premise of the impact hypothesis which is highly disputed within much of the mainstream community. The most widely accepted theory is that this quick freeze was caused by the collapse of the North American ice sheets.

It is interesting that this rapid cooling of the planet has been predicted as the aftermath of a global nuclear exchange in modern times. This condition is known as Nuclear Winter. To suggest that one or more of the Dryas periods might have been brought on by a nuclear holocaust might really stir up some animosity from the mainstream.

It is also quite interesting that the Younger Dryas which ended with another global drop in temperature seems to have been brought on around the same time that Gobekli Tepe appears to have been buried and that the war described in the Mahabharata appears to have been taking place. Perhaps the people living at Gobekli Tepe buried the site in order to avoid being spotted from the air and to avoid getting bombed.

Almost every ancient culture in the world speaks of great catastrophes and of aircraft. Nearly all cultures also seem to have believed that the people emerging from or flying these aircraft were gods. This could be evidence of cargo cultures similar to the one discussed earlier. Although only one of these cultures was included as an example, there were several others. The one used as the example still has a day or period each year to worship a person from that period (World War Two) that is completely unknown, has never been seen and no records of this person appear to exist anywhere.

Historical accounts reveal that when the conquistadors first arrived in America the natives thought they were gods. This turned out to be unfortunate for the natives on this continent but bears a striking resemblance to the legends of more ancient times.

Chapter 15

Ancient Egyptian Aircraft

Below is a photo taken from the internet of a section of the Egyptian temple wall at Abydos.

Some Archeologists have stated that the aircraft depicted here are the result of over writing a previous inscription or simple erosion. Over writing and or erosion might produce one recognizable aircraft but it is unlikely that a group like this would be produced. In this photo alone there are at least four different aircraft designs. One of the aircraft is so futuristic (looks like a flying tank or gunship) that we will have to wait for someone in the future to invent one that resembles it. Another bears a marvelous resemblance some aircraft designed for today's science fiction.

There also appears to be some similarity to the small golden items found in South America.

Another interesting flight related item was found during the excavation of the step pyramid of King Djoser in 1891. When the tomb of Pa-di-Imen who was an official from the third century BC was opened by French archeologists, lying next to this object they found a papyrus bearing the inscription: "I want to fly." No gliders or airplanes had been invented at the time so it was assumed that it was just a bird. It has been called the Saqqara bird since that time. Seventy eight years later in 1969 Egyptologist Dr. Kahlil Messiha was looking at the collection in the Cairo Museum and noticed that the wings are not those of a bird but they do resemble a glider with a modern aerodynamic design. Birds do not have a rudder but this object does. The curved wing is known as reverse dihedral and produces enormous amounts of lift. Testing was done using a scale model five times the size of the object in this photo and revealed that if this bird had a rear stabilizing rudder or elevator it would fly very well. Considering that the tomb was dated to be about four thousand years old it is very possible that the bird may be missing a piece.

The design bears some resemblance to that used on the Concorde aircraft.

Drawings and etchings of aircraft that are too old for any accurate dating are found in stone all over the world. It is somewhat obvious that those who created these images were aware that they would be seen in the distant future. For this phenomena to be so wide spread those ancient people must have seen these vehicles in operation or on the

ground. This one looks like if was probably viewed while it was in flight.

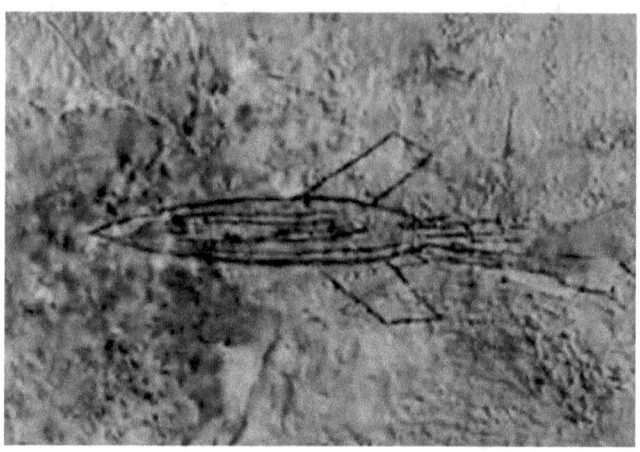

Ancient Rocket Sculpture

This object appears to have been found during an excavation of Toprakkale. Though the site that was excavated dates to about eight hundred BC this artifact may be in fact much older. The Urartian king Sarduri I ruled the Urartu kingdom from eight thirty four to eight twenty eight BC. The city was built using thirty to forty ton blocks of stone and believed by many to have been built long before Sarduri moved his capital city there even though inscriptions claim he built the city. This item was out of context with the rest of the site so it was originally presumed to be a fake. The head is missing but the body and what looks like a pressure suit are definitive. The object has been placed on display at the Istanbul Museum. The detail of this object is extraordinary. It may predate

the site it was taken from and could possibly predate human history.

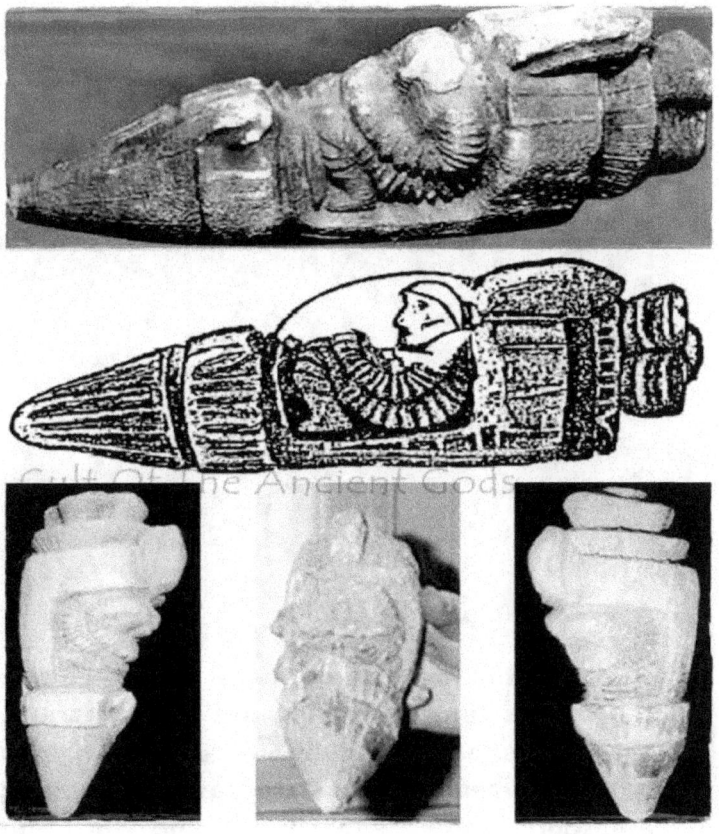

Chapter 16

Electricity

Hathor Temple

Glyphs at the Hathor Temple in Dendera, Egypt

In the opinion of most archeologists, there isn't really anything out of ordinary in these scenes. Explanations of the glyphs have stirred some controversy among amateurs and experts. Some in the mainstream have suggested that this is simply a lotus flower. Where is the resemblance to this flower?

It doesn't take an electrician or an electrical engineer to see the obvious resemblance to a light bulb and the cable running out from the back of the device and into the box where the person sitting on the box has arms positioned in the same position as the arms depicted on the bulb. This would appear to be some unknown power source (much like a small generator or battery) delivering power to light the bulb. It is a well known fact that the interior could not be illuminated with mirrors and there is no residue of any torches or oil used for lighting. For anyone to suggest that this resembles the Lotus Flower does take a great deal of imagination.

Although many different theories have been proposed no acceptable theory for lighting the inside of this temple well enough to carve the relief's has ever satisfied the mainstream or those who believe that this object is a giant light bulb. The Baghdad battery did prove that the ancient Egyptians knew of electricity. Judging from the size of the Baghdad battery and the box viewed in this relief the voltage produced by a battery this size (based on the four volts the Baghdad battery produced and its size difference) would be several times that of the Baghdad battery.

It is also noteworthy that when testing the voltage for this battery the testers used fruit juice. There are many liquids that would have been available to the ancient Egyptians that would have induced a much higher voltage.

The Spinx is one of the world's most unusual monuments and one the most controversial. The mainstream dates this structure at about four to five thousand years old. It is also believed by many that the Spinx is much older that the mainstream will accept. Some believe that the Spinx was built when Egypt had foliage and a much wetter climate and may be as much as ten thousand years older than the current accepted time of its creation. Many theories have been present concerning the monument as well. Also many believe that somewhere under the Spinx a hall of records or great library exist. Ground penetrating radar used over the past twenty years does appear to reveal some rooms or structures under the paw and also some under the Spinx itself have been discovered. Unfortunately no hall of records has been discovered at this time. If this turns out to be true the information contained there could literally change the world and the way the human race currently looks at itself.

Chapter 17

ITEMS FOUND IN ROCK AND COAL

The hammer in the photo below (known as the London Hammer) is made from an alloy of iron which is very modern in technology and is encased in one hundred million year old rock that formed around it.

Detailed research was carried out independently by two different institutes. A number of Australian metallurgists, as well as those working at the respected metallurgic institute Batelle Memorial Laboratory in Columbus, Ohio (USA), took part in the analysis. Electron microscopes were used to examine the structure and composition of the steel.
The results of the examinations were somewhat bewildering. The hammerhead consisted of 96.6 % iron, 2.6 % chlorine, and 0.74% sulfur. Incredibly it is almost entirely solid iron. Other additives or impurities were not detectable. Non-destructive testing

methods showed no evidence of inclusions or irregularities in the hammerhead steel. The even structure suggests that this hard steel was manufactured by some sophisticated technology. The results of the examination are sensational. The modern steel-making process, inevitably leads to carbon or silicon impurities. Steel production without these impurities is almost unthinkable.

No other known ingredients were used for refinement. The high quantity of the chlorine in the fossil hammerhead is remarkable, as well. Chlorine plays no part in modern steel manufacturing. It is not used at all today, so it is probably not possible today to produce steel of this quality. This leads to the question; who manufactured this hammer and when? Based on the standpoint of accepted research and science, it is nearly impossible for this hammer to exist, much less to have ever been manufactured.

The conclusion that it is unlikely that this hammer is a fictitious artifact is all that is left. Much the same has been shown concerning the hammer handle, part of which was in the process of turning to coal when it was found. Two forgery-proof materials with no acceptable scientific explanation, combined in one tool. This could serve to be evidence of a different history of earth and humankind!

The fossil hammer shows still more peculiar features. In breaking open the hammer's original stone enclosure in 1934, the upper edge of the metal head was damaged, leaving a small notch. The inside of the notch revealed a shiny silvery surface. In nearly eighty years the color of the notch has not changed and no traces of rust are perceptible. The mainstream quickly decided that the material encasing the hammer was not really hundred million year old rock and that the hammer was modern (less than a hundred year old) manufacture. However, no one in the mainstream addressed the components of the steel itself. Perhaps overlooking this minor detail was a simple error. Or could be part of some protect my reputation policy?

Machine Parts in Russia

During a cold winter evening a resident of Vladivostok, Russia was lighting a fire to heat his home when he noticed a rail-shaped piece of metal which was embedded in one of the pieces of coal. Being curious he put the piece aside and later took it to some scientists in the Primorye region. Upon examination the object turned out to be gear rail of the type most often used in microscopes and other technical and electronic devices. When tested this rail turned out to be 98% aluminum and 2% magnesium. There is no conventional explanation for this object being inside a three hundred million year old piece of coal.

With no ready explanation available many in the mainstream quickly declared this artifact to be a natural formation. They did not offer any explanation as to how it could be made from "ninety eight percent aluminum", which does not form naturally.

The next photograph shows the object before and after it was removed from the coal.

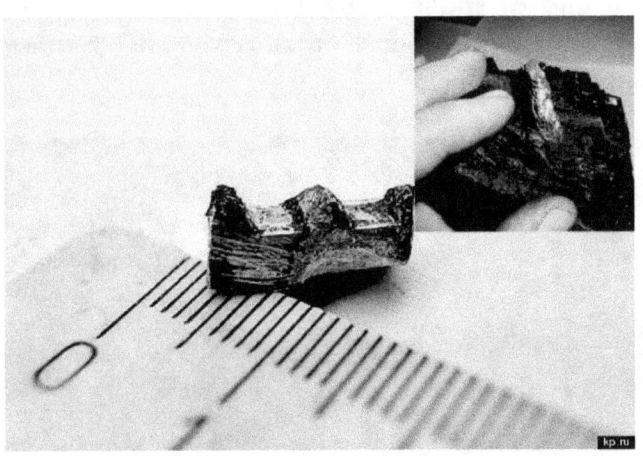

The following photo is of what appears to be gears or parts a mechanical device was found in volcanic rock on the remote Kamchatka Peninsula, 150 miles from the village of Tigil, discovered by archaeologists at the University of St. Petersburg and has been dated to about four hundred million years old. This is located in Kamchatka Krai, Russia no one knows what this device is or was. Currently there has been no subsequent excavation to search for more artifacts or to try and determine what machinery these cogs may have been used in or what their original purpose was.

Just the fact that it is embedded in a four hundred million year old deposit is enough suggest that some incredible information or technology has been lost. It remains difficult to understand why the mainstream does not seem to be interested in investigating discoveries like this when any one of these finds could turn out to be the most profound discovery in human history.

During recent years in Russia's Ural Mountains at depths up to 40 feet thousands of screw like objects made of metal have been found in gravel deposits. Obviously these required a highly sophisticated degree of manufacturing. What these items were used for remains a mystery.

Objects are much smaller that shown in the picture.

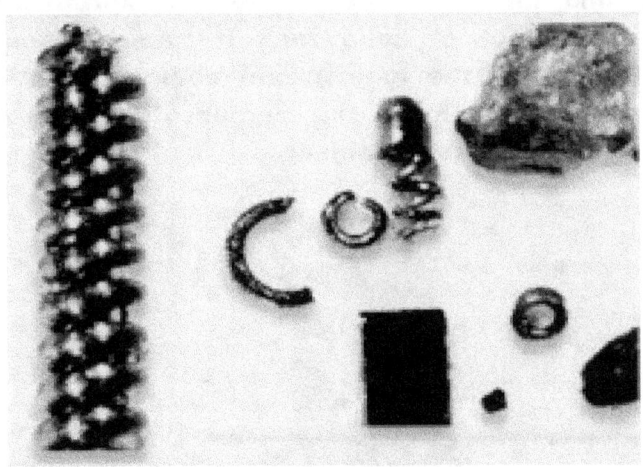

The vase shown here was discovered in Massachusetts in 1851 while blasting in a quarry. This silver-zinc vase has inlays of pure silver to create the vines. The rock level where it was found was dated to be over five hundred million years old. This small vase measures four and one half inches tall and six and a half inches across the bottom. Found in two pieces buried in more that fifteen feet of solid rock it presents and enigma for the mainstream community. We can only imagine the ancient hands that crafted such a fine object.

The iron pot in this picture was found in 1912 when an Oklahoma resident feeding a furnace broke apart a large piece of coal revealing the pot. It has an estimated age of over 300 million years. As with most other objects found in coal the age and obvious technological development varies greatly. This fact points to the possibility of not one but many different rises of civilization. It does not matter if all of the civilizations were human or some other as yet unknown species. After millions of years everything that has not completely decayed would only be preserved in coal or rock or maybe some unknown structure that has not yet been found.

In 1944 a youngster named Newton Anderson found a bell inside a lump of coal that came from a mine close to his home in West Virginia. Newton dropped the chunk of coal and when it broke open he saw the bell embedded inside. This appears to be a brass bell with an iron clapper. The University of Oklahoma analyzed the bell and found that it contained a mixture of metals not known to have been used during accepted human history. To begin with is has an unusual mixture of metals and chemicals copper, zinc, tin, arsenic, iodine, and selenium. This combination produces brass but not of the type that has been used in the history of our current civilization.

Newton was only ten years old when he found this artifact making it very unlikely that he had the skill or the resources to create a fake artifact and also very unlikely that such a young boy would perpetrate a fraud. Nothing resembling this particular style could be found for sale until many years after his find and then only because someone thought a copy of this bell might make a few bucks. Just creating a mold for a piece such as this requires a great deal of work.

111

About twenty miles north of Peoria, Illinois, three men were drilling a well in August of 1870 when one of then noticed a small medal medallion that came up in the drill residue. The medallion was made of an unidentified copper alloy and was about the size and thickness of a U.S. Quarter. The residue came from a depth of around 100 feet. The coin-medallion was uniformly rounded and appeared to have been cut and processed through a rolling mill and the edges still showed machining marks.

Detailed statements were taken from the men drilling the well and several experts of the day examined the coin.

One side had the figure of a woman wearing a crown or head-dress. On the opposite side was another figure, resembling some kind of a crouching animal long, with pointed ears, large eyes and mouth, claw-like arms, and a long tail frayed at the very end. Below and to the left of it there is another animal, bearing a strong resemblance to a horse. Around the outer edges on both sides were some hieroglyphics. Additionally at about 120 feet two copper artifacts came up. These two items appeared to be a hook, and a ring. The depth that produced these items would place them at an age of 150,000 to 250,000 years old. Unfortunately these items vanished while being passed between different people and locations.

All that remains is this sketch of the coin

No excavations of any kind have ever been done in Lawn Ridge to search for more artifacts or to attempt to determine who made these items or when they were actually created. Although there have been many different items found while drilling wells in the United States no major excavation has ever been undertaken. No attempt to discover a history other than that proposed by the mainstream has ever been made.

It appears that once the mainstream decides that history unfolded a certain way there is very little room for new information that contradicts their conclusions unless there is just no way they can ignore it.

Although the evidence is now missing it was handled by many different reputable people of the day. The fact that no one seemed interested in attempting to find out who made the coin or when it was actually made doesn't really point to a science that prides itself on discovery.

Ancient Screw Found In China

The Mazong Mountain area is located on the border of Gansu and Xijiang provinces. It was here that Mr. Zhilin Wang found an unusual black stone about 8 x 7 cm in diameter and weighing about 466 grams and what appeared to be some type of bar tip protruding from one side. When the pear shaped stone was fractured a metal screw was exposed. The screw thread width is consistent from the thick end to the thin end. Groups of geologists and global physicists from the National Land Resources Bureau of Gansu Province, Colored Metal Survey Bureau of Gansu Province, the Institute of Geology and Minerals Research of China Academy, Lanzhou Branch, and the School of Resources and Environment of Lanzhou College have studied this object and as yet have not produced any conventional theory as to how or when it was made. The general consensus is that the material is several million years old.

Foot Prints

The following photo is of a distinct foot print featuring a ribbed sole shoe. Discovered in northern Washington State this print appears to be from a large person and is approximately 16 inches long from heel to toe. Set in granite which Geologists say formed over 1 billion years ago. Either the accepted geology is wrong (a more precise science than ancient history) or someone was wearing shoes a billion years ago. The mainstream quickly labeled this as fake. The point is, according to evolutionary theory no one should have been around early enough to leave a shoe imprint in what is now solid granite. No one should have been around to draw one either. What or who made it and when? Who knows? But the fact that it exist does lead to the conclusion that someone had to be standing there long before this material turned to granite.

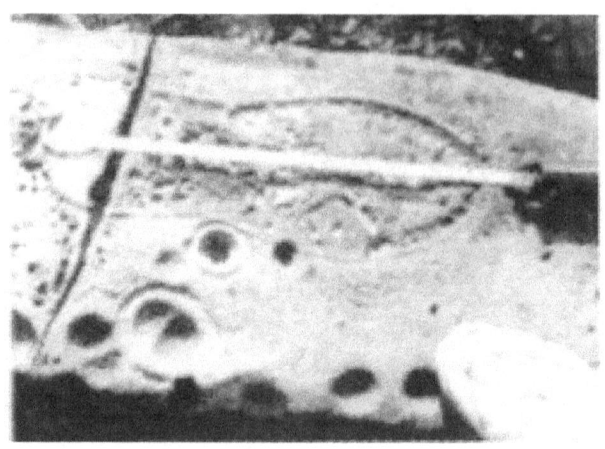

Another instance if foot print evidence was found in Utah in the 1960's. This sandal print is in rock dated between 300 and 500 million years old. There are trilobites embedded in the stone that were crushed by the sandal. Several geologists have examined and confirmed the dating so unless we are willing to discard the science of geology we must accept that someone was wearing sandals 300 million years past. Incredible, considering that mainstream archeology says that nothing even resembling man had evolved at that time. The mainstream seems to want to have things both ways. They use these materials to verify dating of artifacts that they accept but when something shows up they do not like they say the dating is flawed.

During the month of July, 2000 in some North Texas limestone a pristine human footprint was discovered partially mired by a dinosaur footprint. Scientific Verification of

Footprint Authenticity was provided when the fossil was transported to a professional laboratory where 800 X-rays were performed in a CT scan procedure. Laboratory technicians verified compression and distribution features clearly seen in these prints, human and dinosaur. This removed any possibility that the prints were carved or altered.

No human prints should be found during the period that ranged from 145 million to 65 million years ago according to science. According to evolutionary theory the first hominid fossils didn't appear until at least 60 million years later. Finding experts on evolution or paleontology to speak about the discovery proved difficult.

Yet another issue for paleontologists and the rest of mainstream science was found in New Mexico around nineteen eighty seven by paleontologist Jerry MacDonald. Along with many other fossilized foot prints of many extinct species he discovered the footprint of what would appear to be a modern human. The stratum he was studying was dated to the Permian Era. This geologic period extends from about two hundred and fifty million to three hundred million years ago. Long before any of the creatures we know today were supposed to have existed.

The mainstream has only presented the argument that this only looks human. They have not presented any argument that would even slightly contradict its origin. One foot print would convict someone in a court of law but even with two on record both found in recent years the mainstream still will not even consider that their current time frame for the dawn for man or any species similar to man may be wrong. The evidence continues to mount that ancient civilization or perhaps even the human species may be older than the dinosaurs.

Mr. MacDonald wrote an article in nineteen ninety two that was published by the Smithsonian Magazine titled "Petrified Footprints, A Puzzling Parade of Permian Beasts" however he did not discuss this particular footprint in his article. His article can be viewed in the back issue of the

Smithsonian Magazine dated July 1992, Vol. 23, Issue 4, p. 70-79.

As a mainstream paleontologist discussing how a human footprint got imbedded in this material with all the other species of the period must have been a little far out there. Considering the scientific atmosphere of our time it might have even amounted to professional suicide. Still the fact remains that this amazing discovery is on record even though technically it has been shelved and mostly forgotten. Just one more example of how difficult it is for the mainstream to accept anything that does not fit into the current view of how life unfolded on this planet.

Most of our modern archaeological science had been developed in less than two hundred years. It is amazing that when new evidence is presented that may differ from or contradict the current views the mainstream is still reluctant to change those views.

One would think that all science would be open minded enough to at least try and further investigate as many of the anomalies found as possible. Many that have strayed from the accepted norm in the past have become some of the most famous and respected in their field. After all isn't science supposed to be a journey of discovery and a search for the ultimate truth?

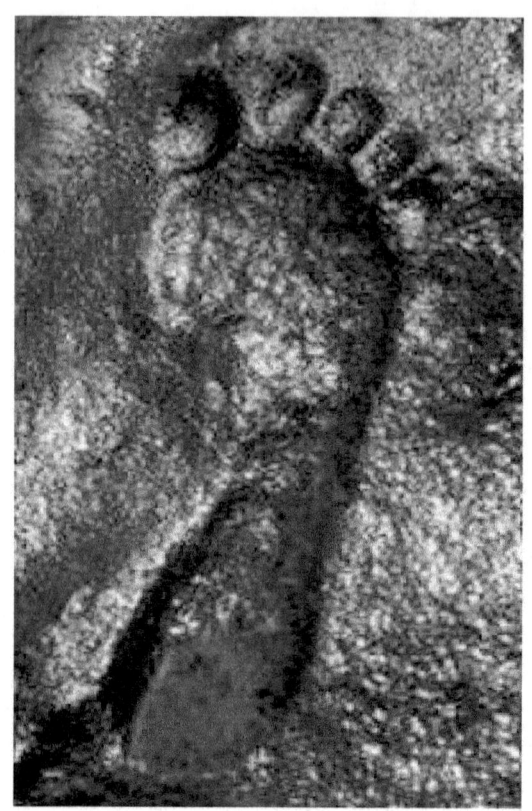

Chapter 18 Underwater Cities

An Underwater City near Cuba

Paullina Zelitzki and Paul Weinzweig, owners of Advanced Digital Communications were conducting sonar surveys off Guanahacabibes Peninsula an area just west of Cuba when their sonar scans appeared to depict symmetrical features aligned to a grid. A second survey a year or two later failed to supply any further information. The images below were produced by the initial sonar scan. If this finding is ever verified it would push the time line for human civilization back to about fifty thousand years and prove that an advanced civilization did indeed exist long before the mainstreams accepted time line.

Satellite imaging at coordinates 31.332525,-24.32375 does appear to show a grid and what could be interpreted as structures. Without further examination we may never know for sure.

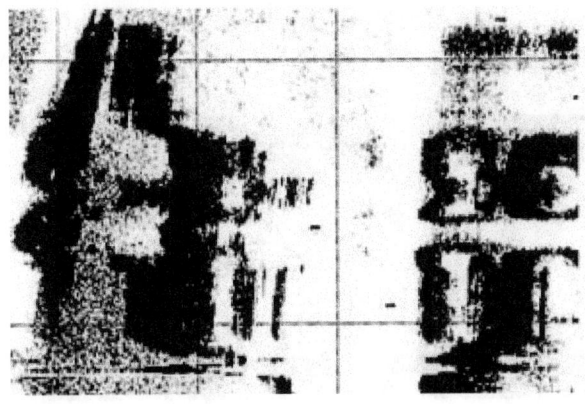

Underwater City in the Gulf of Cambay

Dwarka is an area resting about one hundred and twenty feet below the surface was also discovered by accident in 2001 by Indian oceanographers when side scan sonar returned images of enormous geometrical structures. When the site was investigated Marine archaeologists found human remains, sections of walls, sculpture, beads, and pottery that carbon dated to about seventy five hundred BC. Predating all other known cities in Egypt, Sumeria, and even China by several thousand years.

This city is five miles long and two miles wide which would have rivaled many modern cities in size. Cities of this size are completely unheard of until about forty five hundred years ago. If this city turns out be of Vedic origin the history of India will have to be rewritten. This is nearly double the time line for civilization proposed by the mainstream. It makes one wonder how many accidental discoveries it takes to cause the academics to abandon the five thousand year time line for civilization.

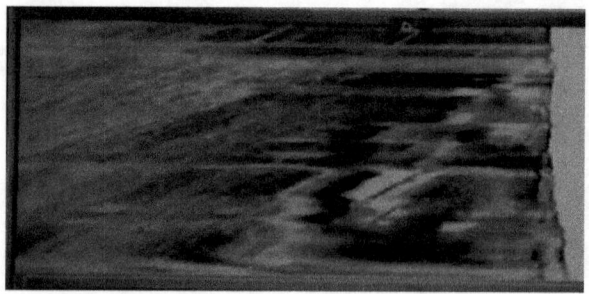

Chapter 19

Incredible Construction

Most people to have heard of Tiahuanaco an ancient city located in the western part of Bolivia which could be in excess of fifteen thousand years old. Truly this is a great wonder of the ancient world.

An ancient building site called Puma Punku is also found here. One of the most amazing ruins in the world is Puma Punku. The stones used for its construction are said to be granite and diorite second only to diamonds in hardness. The most common way to cut granite and diorite is with diamonds. Even with today's technology it would be extremely difficult to cut these stones with the precision achieved at Puma Punku. Granite and diorite are also extremely heavy some of the stones here have an estimated weight of over eight hundred tons. The closest quarry is about ten miles away and there are no trees in the area that could have been used to make sleds or rollers. Even with twentieth century technology it would a monumental task to move just a few of these stones from the quarry to the building site at Puma Punku. That being said it is an astonishing accomplishment for the builders of Puma Punku. Moving the stones was just the beginning. Creating the structure using these colossal stones is nothing short of phenomenal. Even the ruins here rival if not supersede the outstanding accomplishment of building the pyramids of Egypt. When this complex was still standing it must have been practically a religious experience to see it.

Three other structures also stand at Tiahuanaco, The Akapana Pyramid, The Kalasasaya Platform, and The Subterranean Temple. All of these defy logic and cause the imagination to run wild. If the human race could solve the mysteries of this one location it would not only rewrite history but give us have some idea of the knowledge that has been lost.

It must have taken a monstrous event to splay these stones about destroying the original structure. Perhaps something like the Tunguska event in 1908 when the energy released was somewhere near a thousand times the atomic bombs dropped on Japan near the end of world war two. Or something man made that had such great power.

These stones were finely cut. The cuts are perfectly straight. The holes cored into these stones are all of equal depth and absolutely perfect. The interlocking stones were cut to be almost identical and to such a close degree that they could be randomly assembled. Details such as this display an advanced knowledge of geometry, excellent planning and probably mass production.

One can not say enough about the quality of this work. No words can put into context the true greatness of these builders or the civilization that could contemplate such a task.

Even using current building practices and all available technology creating a site like this would be a monumental task if it is even possible. The transport of the stones alone might prove impossible for today's technology.

These ancient people were obviously highly advanced. Then at some point a catastrophic event wiped them out, perhaps the same event that ruined the complex itself. One might note that the object that exploded above Tunguska leveled trees across an area covering eight hundred and thirty square miles and the shock wave would have measured 5.0 on the Richter scale. A blast of this size could definitely wipe out a large population living within the general area. We may never know for sure what happened to these people or who they were but the legacy they left behind has endured for thousands of years. And no doubt will continue to baffle science for many years to come.

Puma Punku is believed to be over seventeen thousand years old although no absolute or accepted date for the building of this site is available so it could be much older

than currently estimated, however it is widely accepted that it most likely predates human history by several thousand years.

Sacsayhuamán Walls

Another amazing site is located in Peru. Sacsayhuamán is located near the city of Cusco. This site is constructed of all irregular cut stones and does not have two stones that are cut the same in the entire structure and yet a single knife blade will not fit between the joints. The entire structure has a slight inward slope giving it great stability

No one knows if this was a place of worship, some kind of safe compound (like a fort) or if it was just some kind of monument in the shape of the puma to signal the gods that Inca were there.

If this structure had not been highly resistant to earthquakes it would have disappeared long ago. The technology that was used in building this site has also been lost over time.

Derinkuyu

In nineteen sixty three a discovery made in the Derinkuyu district in Nevşehir Province, Turkey shocked the mainstream community. This accidental discovery turned out to be an underground city. Its eighteen stories could house as many as twenty thousand people. It was complete with room for livestock and food stores it also had a water supply that would have maintained the entire population. A large chamber in the structure may have been used as a religious center and or perhaps a school or meeting place. Much like the town churches of the past few centuries were used for multiple functions. This amazing underground complex has doors that could only be opened or closed from the inside. It only required one person to open and closed the heavy doors.

Although there are other known underground cities, Derinkuyu is by far the largest one found to date. There is at least one known tunnel that connects this complex to another one that is approximately five miles away. The mainstream labeled this city to be about twenty eight hundred years old and have found artifacts from that period up through the Byzantine era. However, there are many that believe the site could be in excess of ten thousand years old. That dating would also put this sites construction at about the same time as the legendary catastrophes described by many ancient cultures.

Obviously this enormous bunker was not carved out by people using primitive picks. Once again this appears to prove that ancient man was in possession of some kind of technology not known or understood today. It took a great deal of planning and an amazing understanding of geology to construct such and underground dwelling that would survive thousands of years. Excavation continues at this site even after fifty years. Who knows what may be found next?

Ollantaytambo

Ollantaytambo, Peru sits almost ten thousand feet above sea level. Many of the stones used in this ancient construction weighed several hundred tons and were cut with a precision that could only be matched with the laser cutting technology of today. Stones are fitted so well that a sheet of paper can not fit between them even thousands of years after the construction ended. It is easily seen how well the stone is fit together by looking at the wall of six monoliths.

Mainstream archeology dates this site to the sixteenth century, mostly because of the reconstruction efforts by Emperor Pachacuti after the destruction of the original city when he conquered it. These ruins are believed by many to be much older and may even predate the Inca.

Chapter 20

Ancient Maps

Conventional science says that the ice-cap covering the Antarctic is millions of years old. Although there have been some unsubstantiated claims that the northern portion of that continent was ice free as recently as twelve thousand years ago.

Piri Reis, (a famous admiral of the Turkish fleet in the sixteenth century) drew a map that showed the land masses as well as the true coastline under the ice in 1513. Notes on the admiral's map indicate that he used data from several source maps located in the Imperial Library of Constantinople. In his notes he indicates that some of the source material dated back to the fourth century BC or earlier. Amazingly the Antarctic would not actually be discovered for another three hundred years and we would not be able to map the coastline or the underlying land masses for nearly another hundred and fifty years or the nineteen sixties. The Antarctic is supposed to have been covered with two miles of ice for the past two million years. So somehow the mainstream will need to explain how an undiscovered region that had two miles of ice on it could possibly be drawn in such great detail without aerial surveys, penetrating radar imaging or and other know technology that could map below that much material.

It would be impossible for the mainstream to label these maps were fake because they were verified in nineteen fifty eight when our own people mapped the region and produced a perfect match. After all one can not fake a map three hundred years before the locations is discovered and another hundred years before the information is known and then have it verified as accurate.

Prof. Charles H. Hapgood of Keene College sent a request to the U. S. Air Force to evaluate the Piri Reis map. The Air Force responded with the following letter.

TO: Prof Charles H. Hapgood, Keene College
Dear Professor Hapgood,
Your request of evaluation of certain unusual features of the Piri Reis Antarctica map of 1513 by this organization has been reviewed.. The claim that the lower part of the map portrays the Princess Martha Coast of Queen Maud Land, Antarctic, and the Palmer Peninsular, is reasonable. We find that this is the most logical and in all probability the correct interpretation of the map. The geographical detail shown in the lower part of the map agrees very remarkably with the results of the seismic profile made across the top of the ice-cap by the Swedish-British Antarctic Expedition of 1949.This indicates the coastline had been mapped before it was covered by the ice-cap. This part of

Antarctica ice free. The ice-cap in this region is now about a mile thick. We have no idea how the data on this map can be reconciled with the supposed state of geographical knowledge in 1513.
Harold Z. Ohlmeyer Lt. Colonel, USAF Commander

The Piri Reis map was later sent to the U.S. Navy Hydro graphic Bureau in 1953. A grid was made and when the map was transferred onto a globe it was a perfect match. Authorities on ancient maps stated that the only way to draw a map of such accuracy was to do aerial surveying. Some details from this map were actually used to make corrections to modern maps of 1953. Due to the fact it would be another five years before the information from the Antarctic portion of the map could be verified.

To draw such maps, the authors had to know about spheroid trigonometry, the curvature of the earth, longitude and latitude, and methods of projection, this information not known until the latter half of the eighteenth century.

The early map makers knew that the Earth was round and had knowledge of its true circumference to within 50 miles!

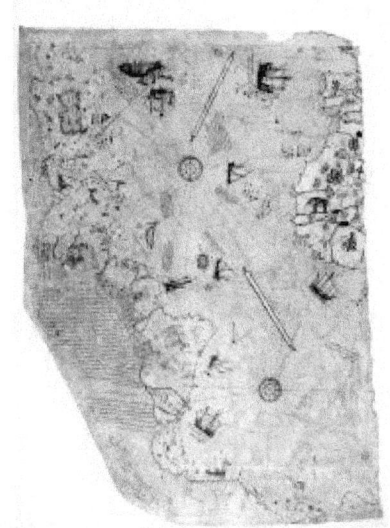

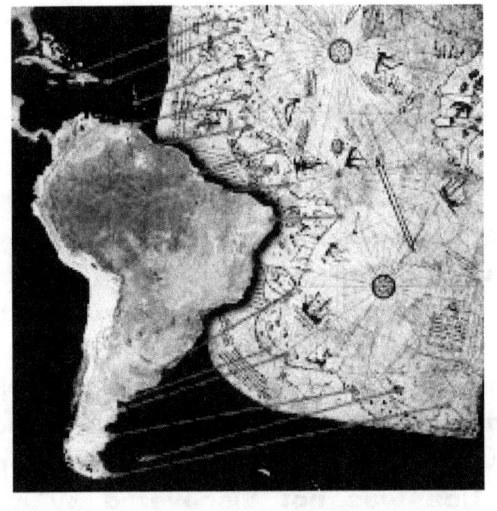

A map drawn in the year 1532 known as the Oronteus Finaeus map also shows an ice free Antarctica with flowing mountains, drainage patterns, a clean coastline and rivers. This map also shows Greenland as two separate islands, only hundreds later was it confirmed that an ice cap does indeed join the two islands that make up Greenland. Antarctica was not discovered by modern man until 1820 nearly three hundred years later.

World Maps with such intricate detail could not be produced by modern man until the latter half of the twentieth century.

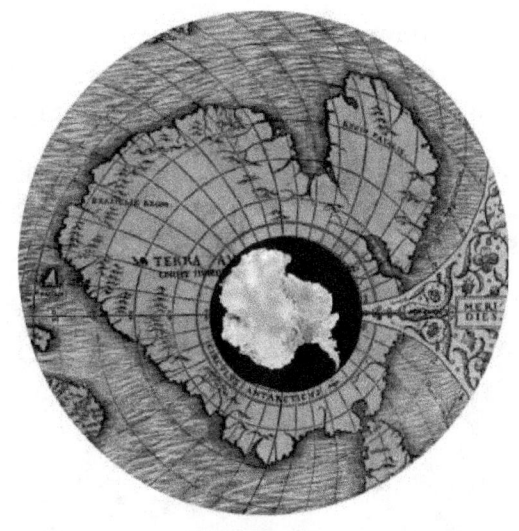

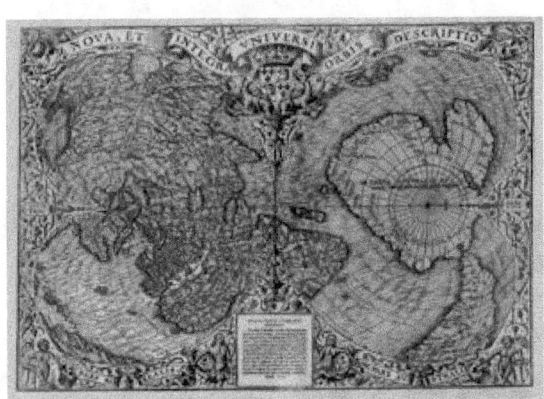

Yet another map with this great quality and excellent a detail was drawn by Hadji Ahmed, in 1559 and includes the land bridge that joined Alaska and Siberia before the sea levels rose at the end of the last ice age. There is no doubt that all of these maps were copied using older maps for source material. Too bad that none of the original source maps have survived to the present day.

It is amazing that in an era when conventional science believed the earth was flat and that if one sailed too far they would drop off the edge that maps like these were not only drawn but survived (heresy was a hanging or burning offense in those days if you were lucky).

Chapter 21

Engraved Stones

The Ica Stones of Peru

These andesite stones are engraved with images of humans riding dinosaurs, many animals, highly advanced medical procedures, and technological instruments such as telescopes and several different types of aircraft. A total of about 15,000 stones have been recovered. Most of which are stored in the Ica Stones Museum in Peru.

These ancient stones are covered with a varnish over them that appears' to have accumulated over thousands of years. Lighter colored lines appear when the outer layer of varnish is removed. Upon close examination it was found the carvings also have some varnish on them indicating that they are also of ancient origin.

The farmer that originally found these stones was selling them until some archaeologists got interested and got him arrested for selling national treasures. At point that a confession was coerced from the farmer claiming that he did the engravings the farmer had little choice it was simply confess or spend the rest of his life in prison. Out of fear of being arrested and imprisoned he did not recant the confession until near the end of his life when he had nothing to loose. The fact is that one man could not carve so many stones large and small in one lifetime. There is also no way to add the accumulation of ancient varnish that covered the stones.

Jesuit missionary Padre Simón came to the Americas with Pizarro to Peru in fifteen twenty five. The Padre gathered some of these stones and shipped samples to Spain in fifteen sixty two (about four hundred years before the farmer of the nineteen fifties was born). Some of the stones dug up in the fifties and sixties have images of an Apatosaurus with the correct head, but it was not until nineteen seventy nine that it was revealed in the scientific community that paleontologists had been using the wrong skull when assembling the Brontosaurus fossils. Another correction to modern information from a source that the mainstream would like to label as fake..

Some stones depict humans battling dinosaurs and riding them. When the first of these stones were discovered in the sixteenth century no one even knew that dinosaurs ever existed.

These stones also have depictions of surgical procedures that were unknown at the time they were found. Heart transplants did not occur in the modern world until the latter half of the twentieth century. Brain surgery was in its infancy at best. Kidney transplants were still unknown, also a development of the latter half of the last century. There are also host of surgical procedures that are still unidentified. No doubt many of the unidentified procedures will be available in our near future. How the mainstream can still label such artifacts as fake is a matter of contention. We may never know the truth or reason behind that.

Interesting that a native seems to be looking at the stars through a telescope (pictured below) that according to the mainstream was not available until Galileo Galilei invented it in the seventeenth century. Although not available to our civilization it would appear that it was available to the ancient people that created the art on the stones. Perhaps there is a possibility that after a nuclear holocaust the only material available to attempt to salvage information from an advanced civilization was stone. Imagine if we destroyed our civilization tomorrow. It would not be long before the paper was gone and our technology would probably disappear sooner that the paper. Past civilizations may not have had as many citizens as ours, probably wouldn't have taken up as much space as we do, but their existence is a probability.

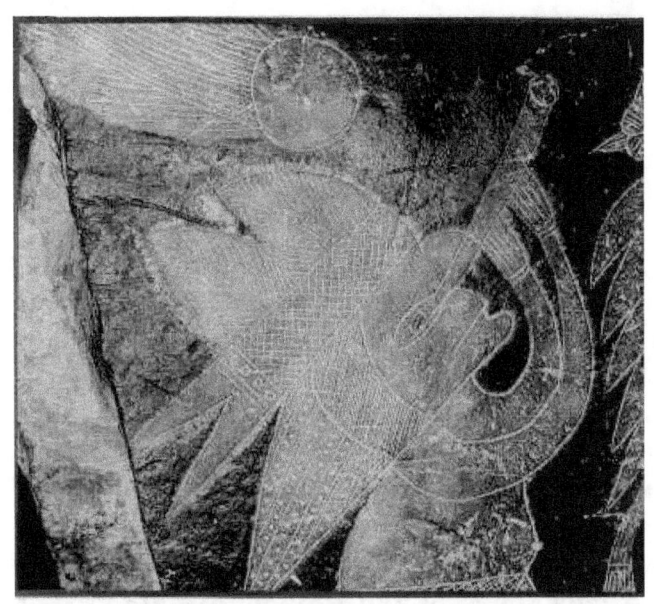

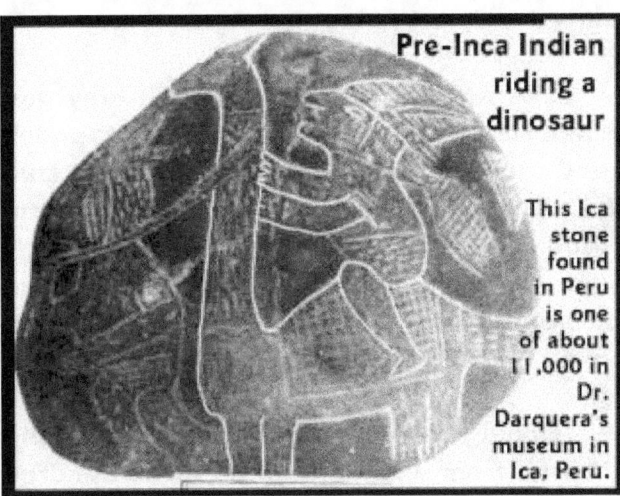

Pre-Inca Indian riding a dinosaur

This Ica stone found in Peru is one of about 11,000 in Dr. Darquera's museum in Ica, Peru.

Chapter 22

Things brought up by well drillers

The Nampa Doll

In August eighteen eighty nine three men were drilling a well in Nampa, Idaho. At a depth three hundred and twenty feet their steam pump brought up piece of brownish clay about one and a half inches long that was shaped like a human being and on closer inspection turned out to be a small doll. They had been drilling through clay, sand and a fifteen foot layer of lava rock deposited by the lava flows of the Columbia Plateau that occurred before the last Ice Age.

The doll is composed of half clay and half quartz. Experts that examined the doll agreed that it was the product of a true artist. Although badly worn by extreme time the dolls features were still visible. Markings around the neck and wrist appeared to have once been a representation of jewelry. Much the way figurines are manufactured today.

Geological estimates for the depth the doll came from was as high as three hundred thousand years. The report of the presence of iron on its surface would indicate that it is of ancient origin but no definitive proof of the actual age has ever surfaced. As is always the case many varied stories surrounding the object have been offered to both verify and disprove the antiquity of the object. Dr. G.F. Wright of the Boston Society of Natural History examined the doll and conducted interviews to verify the depth it came from and in the end was convinced that it was an authentic artifact. This item can be viewed on display at the Idaho State Historical Society in Boise.

A similar doll-like figure was discovered sometime before 1880 near Marlboro in Stark County, Ohio, also by workmen drilling a well and also pulled up bay the drill pump. This one was made of a black marble not indigenous to Ohio and stood about six inches tall and the features bore a strong resemblance to the Nampa doll. Unfortunately a photograph of this object could not be located.

No excavations have ever been done in either of these areas to search for more objects.

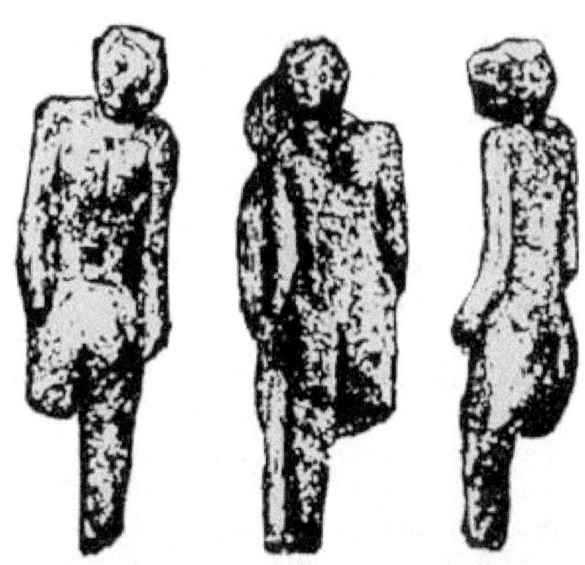

During the years seventeen eighty six and seventeen eighty eight, Aixen -Provence, France was rebuilding their Palace of Justice. A local quarry was the source being used to provide the limestone necessary to complete the job. The rock strata in the quarry were separated by layers of sand. Workers would remove a layer of rock then a layer of sand to get to the next layer of rock. After removing eleven layers of rock they began removing another layer of sand when they started finding stumps of stone pillars and fragments of half worked rock just like the rock they were working. Further down they found coins and pieces of petrified wooden tools and fragments of tools resembling the very ones they were using to work the stone. As they continued to remove the layer of sand they came across a petrified wooden plank resembling the quarry man's plank they used. Although the plank was broken into pieces it was obviously a wooden plank. After reassembling the plank they observed that it was worn on the edges just like the one they were using. No one could figure out how these items became buried under three hundred million years of stone and sand. This find was discussed by the locals for a few years and then forgotten (or maybe the interest was lost since they did not have any answers) until it resurfaced in the early eighteen hundreds. Much like artifacts found today this event was largely ignored by the scientific community of the time.

This event was reported by the American Journal of Science and Arts in eighteen twenty on pages one forty five and one forty six. The quote shown on these two pages is from Count Bournon in his work of Mineralogy. The original account of this incident written by the Count was in French.

Chapter 23

The Worlds Largest Effigy Mound

Great Serpent Mound

Here we have another effigy that has been disputed by modern science ever since it was discovered. Even the time line for its discovery is somewhat skewed. Checking Wikipedia we find this object was supposed to have been first reported by Smithsonian surveyors in eighteen forty eight (nineteenth century). Later in the Wikipedia article it states that an eighteenth century missionary reported that the local Indians told him that the Allegheny people built it in ancient times.

Then we have the mainstream with carbon dating varying any where from the tenth century AD to the twenty ninth century

BC. It seems that every time they test they get a different date depending on where the get their material for carbon dating. Proving the truth is they have no valid data for who built it or when. There is no record of an Allegheny ancient or otherwise living in the area, except of course for the local Indian legends.

Then there is the fact that the mound is best observed from the air. We can tell what it is from the ground but to truly appreciate its beauty and complexity we need to fly over it. The mound totals an average of three feet high and thirteen hundred and seventy feet long and at the end with the head appears to be swallowing an egg. Although some argue that it is a depiction of a solar eclipse.

Since the best view is from the air we can assume that that is probably where it was intended to be viewed from, perhaps the same way as those built by modern cargo cultures. As with most things that do not fit their accepted view the mainstream completely disregards the Indian belief that the Allegheny built the mound in ancient times.

The mound is built at the edge of a two hundred and fifty million year old meteorite impact crater. Of course those in the mainstream disagree on the intent of those who built the mound as to whether it intentional or not. Again proving their general unwillingness to give credit to ancient people for knowing things they were not supposed to by modern academic standards. Many effigy mounds were built in North America and around the world. There are an estimated fifteen to twenty thousand mounds built in Wisconsin alone with about four thousand left in existence today. However, Great Serpent Mound located in Ohio is the largest know such effigy in the world.

Chapter 24

The Genetic Disk

One of the most astounding artifacts known is the genetic disk. The disk has carvings of the various stages of development of a fetus from conception to birth. If the depictions stopped there it would be incredible, but they do not. There are depictions of cell division that can only be seen under a microscope. No wonder that the mainstream cried fraud when this object surfaced.

The object is made of Lydite which is nearly as hard as granite but it is a very flaky material. Some believe that we would not be able to duplicate such and artifact today even with our modern technology. It is between eight and a half and eleven inches in diameter (depending on which report we read) and weighs around to four and a half pounds. The object has been estimated to be a much as six hundred thousand years.

Depictions of biological process can not be created by accident. Someone knew these processes when the stone was created. Most of this information was unknown to our modern world only being discovered over the past two hundred years or so. Once again the ancients seem to have a lot more information than they were supposed to be capable of.

As these tantalizing artifacts continue to be found it is only a matter of time before out ideas of the past and our history will have to be rewritten in its entirety. It would appear that our modern science may not be new science but a rediscovery of science that was know long ago.

Chapter 25

Ancient Tile Floor

According to the mainstream the North American continent did not have any people living on it until sometime between twelve and fifteen thousand years ago when they migrated across a land bridge that connected the continents of Eurasia and America. The land bridge eventually collapsed leaving what is now known as the Bering Strait.

In nineteen sixty nine while working on a street extension in Oklahoma City after workman cut through a rock shelf and about three feed down they found an inlaid tile floor. The Edmond Booster edition of July third nineteen sixty nine reported that an Oklahoma City geologist by the name of Durwood Pate stated that he was sure the floor was man made because of the regular placement of the tiles. Just a few days earlier in the Tulsa World edition of June twenty ninth nineteen sixty nine quoted another Geologist Delbert Smith who was also president of the Oklahoma Seismograph Company as saying "There is no question about it. It had been laid there, but I have no idea by whom."

There were several thousand square feet of the tiles placed in intersecting parallel lines to for a diamond pattern pointing east.

Post hole were found at regular intervals suggesting structures for multiple habitation.

The best geological estimates for the age of this file floor were two hundred thousand years or more. Maybe the mainstream should rethink their twelve thousand year time line.

This is just one of the anomalies that have been found in America that dispute the mainstream theories of ancient colonization and migration. Yet nearly fifty years after this discovery they have not revised estimate for how long people have lived in the Americas.

Chapter 26

When someone not involved in mainstream science finds something out of the ordinary it is usually difficult to get anyone in the mainstream to evaluate the item. This appears to be the case whit John J. Williams (CEO Consumertronics, P.O. Box 23097, Albuquerque, NM 87192) who made an unusual discovery while hiking in nineteen ninety eight. Somehow his attention was drawn to what appeared to be an ordinary rock. Embedded in the rock is what appears to be some type of electrical device.

For the past fifteen years Mr. Williams has attempted to get someone from the mainstream to examine the rock and the embedded object with very little success. He has literally dared the mainstream to prove it is a fake or authenticate it. Unfortunately most of those in the mainstream have not responded well. In their smugness they have labeled the object as fake without even examining it. Mr. Williams has websites with photographs of the rock complete with x rays. Websites can be viewed at the following location on the internet http://www.consumertronics.net/petradox2.htm Mr. Williams has also stated that the object is available for examination to any researcher or reporter with the accredited skills. He has recently offered the object for sale.

At least one comment found on the internet was from someone stating that they would have expected patina on the prongs due to the extreme age implied.

Evidently the person making the comment did not read Mr. Williams article where he stated that he spent several hours cleaning the exterior coatings away in order to see what lay underneath. If the author of the statement was looking for rust he or she might consider that there are many metals that will not rust no matter how old they are.

It is a great shame that the mainstream is so concerned about reputation or perhaps its just peer opinion that they will not investigate claims such as this or that they will make an uninformed assumption about such objects simply because it doesn't fit their paradigm.

Chapter 27

Found in North America

The Malachite People

Another instance of the mainstream following the current dogma can be seen in the example of a bulldozer operator in Utah that discovered ten sets of human bones beneath fifty to a hundred feet of sandstone in the mid nineteen nineties.

The mainstream states that this is most likely a recent or near recent entombment citing a possible mining accident or intentional burial. The facts that there were no indications of a working mine or that native Americans and no one else known for that mater have ever buried their dead in fifty or more feet of sandstone were completely ignored. They also ignore that there was no apparent trauma that would be present if something like a mine cave in had buried these people.

Dating for the level in the limestone was estimated to be about one hundred and fifty million years old. There were remains of small children which would not be working in a mine (also ignored by the mainstream). The positions of the bodies would indicate that they may have been caught in a flood and possibly drowned. There were no tunnels found at the site or any other evidence to indicate that these people got any other way.

As part of their consensus that this was intentional or accidental entombment the mainstream cites that the skeletal remains appear to be of modern humans. Since the mainstream discounts the possibility that man has been around lounger than their current theories proclaim any other explanation or theory about how these people came to be in this place was quickly discounted.

Ancient Wall

In Bradley County, Tennessee on the Hooper farm in eighteen ninety one a wall with some for of ancient writing on it was accidentally discovered. The wall was mostly buried with taller stones rising above the surface every twenty to thirty feet. The Smithsonian sponsored a dig at the site which uncovered a red sandstone wall about eight feet high and two feet thick. The wall follows a ridge to the Tennessee River where continues down and under the river. The wall was estimated to be about a million years old. The entire surface of the western face had been finished and inscribed with the ancient language.

Interpretations of the language (from the mainstream) on the wall ranged from some form of ancient Hebrew to markings left by worms or insects. One thing is certain worms and insects do not draw pictures or repeat patterns in non specific intervals. The wall has been dug out and removed. No one knows who did the work or where the wall is now.

The information in this work was compiled from research on the internet and many book publications. The opinions are strictly to present an alternative view that allows the reader to decide what they want to believe. This is not the first work to present a catastrophic view nor will it be the last. New information and new artifacts turn up almost every day. Eventually someone will find out the truth. It could be you.

www.ingramcontent.com/pod-product-compliance
Lightning Source LLC
Chambersburg PA
CBHW051702170526
45167CB00002B/498